GIS Tutorial for
Humanitarian Assistance

Firoz Verjee

ESRI PRESS
REDLANDS, CALIFORNIA

Ask for ESRI Press titles at your local bookstore or order by calling 800-447-9778. You can also shop online at www.esri.com/esripress. Outside the United States, contact your local ESRI distributor.

ESRI Press titles are distributed to the trade by the following:

In North America:
Ingram Publisher Services
Toll-free telephone: 800-648-3104
Toll-free fax: 800-838-1149
E-mail: customerservice@ingrampublisherservices.com

In the United Kingdom, Europe, Middle East and Africa, Asia, and Australia:
Eurospan Group
3 Henrietta Street
London WC2E 8LU
United Kingdom
Telephone: 44(0) 1767 604972
Fax: 44(0) 1767 601640
E-mail: eurospan@turpin-distribution.com

Cover: Satellite imagery courtesy of GeoEye.

Contents

Preface

In early 2002 I was hired by a small nongovernmental organization (NGO) called FOCUS Humanitarian Assistance, to establish its international GIS program. My first impulse was to study how other, larger humanitarian agencies exploited GIS, so that I could mimic industry standards. I simply didn't have enough time or money to relearn their lessons, and I hoped to follow an established doctrine in the field of humanitarian GIS. It was quite a surprise, therefore, to discover that very few humanitarian agencies operated formal GIS programs, and almost none employed GIS for analytical purposes. The application of GIS was primarily *ad hoc* and cartographic in nature: United Nations, donor, and NGO staff using whatever shapefiles and images they had to produce eclectic maps illustrating Who was doing What and Where (the notorious "3Ws").

Carmelle Terborgh, the manager of ESRI's humanitarian and international development programs at that time, assured me that my observations were not inaccurate. Establishing a formal GIS program at a humanitarian NGO was indeed unusual. So, with the encouragement and support of Jack Dangermond and the late Bill Wood (U.S. State Department), I began to build a modest but systematic GIS program at FOCUS, which at that time was heavily involved in repatriating Afghan refugees back to their homeland, from camps along the Pakistani border.

Since then many things have changed: Internet access has tremendously improved; GPS receivers and computers are more affordable; and high-resolution commercial satellite imagery has become ubiquitous. In addition, GIS software has become much easier to use, and geovisualization tools can now be downloaded for free. But there still remains a remarkable lack of doctrine in how to properly exploit geographic information during humanitarian emergencies. *The purpose of this book is to provide the core skills necessary to realize the full potential of GIS in humanitarian assistance.*

If you have not already had basic training in ArcGIS, you are strongly advised to complete *GIS Tutorial: Workbook for ArcView 9*, third edition, updated for ArcGIS 9.3 (ESRI Press) or a similar course before beginning this tutorial, since it advances rapidly from novice to intermediate and advanced methods. Also, note that the introductory material at the beginning of each chapter is not meant to be comprehensive. I strongly encourage you to consult other available references, especially if you are working outside a classroom setting. (Use the online resources mentioned in the book to get started.)

GIS Tutorial for Humanitarian Assistance includes a DVD containing a 180-day trial version of fully functioning ArcGIS Desktop 9.3.1 software (ArcEditor license level) and a DVD of data for working through the exercises in each chapter. (See "Notes for Instructors" for how this tutorial can be used in a classroom setting.) A data dictionary appears in each chapter, listing each dataset's layers or attributes by file name, along with their descriptions. The GIS professional begins by studying a dataset's origins and contents before using it. In the same way, the student should review the applicable data dictionary before beginning each chapter's exercises.

The first two chapters of this tutorial provide an overview of the field of humanitarian GIS and its cartographic applications. Chapter 1 offers a basic description of GIS and its role in humanitarian assistance, and reviews the ArcMap software interface. In your first lab you will create a simple map for a region of Pakistan that suffered an earthquake in 2008. Chapter 2 introduces basic data loading and display skills, using data from the World Food Programme's food security assessments in Timor-Leste. You will add, manipulate, and symbolize feature layers and identify regions of greatest and least per-capita food sufficiency.

The next three chapters focus on spatial data management—a major weakness among humanitarian GIS users today. Chapter 3 describes the various types of data that can be managed within a GIS and reviews the key principles of projection systems, geodatabases, and metadata creation. Chapter 4 explains how to populate a geodatabase with useful data. The chapter emphasizes the importance of knowing how to find and import existing spatial data, and how to create your own custom data, in order to exploit GIS most effectively. In chapter 5, you will experience for yourself the advantages of using industry data models to design a geodatabase for your organization.

The last four chapters explore the analytical applications of GIS. Chapter 6 introduces methods used during the Kosovo War to reduce the threat posed by land mines and related hazards. Chapter 7 demonstrates the potential for transportation network modeling in Ethiopia to determine optimal locations for warehouses and delivery routes for humanitarian aid. Chapter 8 explores of the use of GIS for selecting IDP (internally displaced persons) camp locations in northern Uganda. Terrain, land use, accessibility, and proximity to hazards are some of the criteria used to determine site suitability. Chapter 9 then uses an actual camp operation in northern Uganda to show how humanitarian services can be improved using guidelines provided by the Sphere Charter and the *Handbook for Emergencies* from the UNHCR (United Nations High Commissioner for Refugees).

Completing all nine chapters of this tutorial will not only prepare you for most routine aspects of humanitarian GIS, it will also teach you how to perform many advanced GIS applications that offer valuable insight to decision makers during an emergency. It is worth remembering that the field of *humanitarian GIS* is only about a decade old—your mastery of the skills provided by this book positions you at the forefront of this field.

Notes for instructors

The objective of this book is to demonstrate how GIS can be feasibly applied during humanitarian emergencies. This is the first book that specifically teaches the humanitarian applications of GIS: it builds upon the recent experience of leading practitioners from around the world and establishes some basic doctrine for the analytical applications of ArcGIS software during humanitarian operations.

Instructors should ensure that students have already attained basic proficiency in ArcGIS before beginning this book. Suitable resources for preparation include the GIS Tutorial series and *Getting to Know ArcGIS Desktop.*

Tutorial structure

Because there are many excellent books and courses that teach general GIS theory and practice, this tutorial makes no attempt to replicate information that is already widely available to teachers and students around the world. It concentrates on the humanitarian applications of GIS, especially its analytical applications, a field that has so far received relatively little attention by educators and academic publishers.

This tutorial is designed as a complete training system for novice GIS users interested in attaining advanced skills in humanitarian GIS, and instructors are encouraged to proceed through each of its chapters in sequence. It consists of nine chapters of increasing difficulty.

Chapter 1 introduces how GIS can be applied to humanitarian operations and summarizes the various categories of analytical applications of GIS. It also includes a "refresher" on the ArcGIS Desktop user interface and a review of rudimentary skills, such as manipulating map layers and creating an orientation map.

Chapter 2 explores how to map thematic data within the context of the World Food Programme's (WFP) activities in the Southeast Asian country of Timor-Leste (East Timor). Students will build essential cartographic skills, learn how to load, join, and display tabular data, and present their work to decision makers.

Chapter 3 delves into a neglected aspect of humanitarian GIS: the proper management of spatial data. Within the framework of the file geodatabase, students will build their own geodatabase, learn how to manage data with geographic and map projection systems, and then create metadata in accordance to widely recognized standards.

Chapter 4 builds upon the previous chapter using MapAction's experience during the 2007 Ghana Floods. It helps students learn how to be resourceful in populating their geodatabase quickly and cheaply, given the reality that most humanitarian operations occur in regions with poor spatial data infrastructure. With the completion of this chapter, students will have attained important skills in building, maintaining, sharing, and exploiting spatial data using ArcGIS.

Chapter 5 introduces data modeling operations using the UNSDI Transport data model. Students will download and modify an industry-standard data model so that it meets their requirements, and then load that model with their own data for analysis. Although data models are not yet widely adopted by the humanitarian community, this lesson builds upon the skills developed in chapter 4 and is a recommended prerequisite for chapter 7.

Chapter 6 takes the student back to hazardous conditions facing the humanitarian community in the aftermath of the Balkan War in the late 1990s. Using spatial transformations—buffers, intersects, unions, and so forth—students will develop expertise in hazard management and improve their ability to produce decisive information for stakeholders.

Chapter 7 describes how to use the data model from chapter 5 to perform a series of logistics planning exercises for the WFP's Targeted Supplementary Feeding Programme in Ethiopia. The student will use the ArcGIS extension Network Analyst to calculate distances and travel times between various locations and to analyze service areas and routing strategies between several food warehouses and hundreds of delivery points throughout the country.

Chapter 8 continues to advance core analytical skills using the experience of IDP (internally displaced persons) campsite selection in northern Uganda. Students will use ArcToolbox and another ArcGIS extension, Spatial Analyst, to analyze a variety of raster and vector datasets to determine suitable locations for a new IDP camp. They will then employ ModelBuilder to automate the site selection process.

Chapter 9 builds upon the previous chapter by examining the humanitarian services within a real camp in northern Uganda. Students will digitize existing camp infrastructure using high-resolution satellite imagery, analyze population density, and then generate a camp-level address system. They will then apply several forms of geostatistical analysis to improve services within the camp. This chapter demonstrates the potential for using GIS to assess compliance with the humanitarian guidelines provided by the Sphere Charter and the *Handbook for Emergencies* published by UNHCR, the United Nation's Refugee Agency.

Pedagogical approach

The above chapters can be used independently to teach specific skills or bundled to create customized training programs. The following three bundles provides a useful strategy for specialized training programs:

Bundle A—Introduction to humanitarian GIS (chapters 1, 2, 3)

Bundle B—Basics of spatial data management (chapters 3, 4, 5)

Bundle C—Analytical applications of humanitarian GIS (chapters 5, 6, 7, 8, 9)

Every effort, however, should be made to teach the entire tutorial in conjunction with supplemental training in the field and in the theory of humanitarian GIS. Holistic approaches have the greatest impact on student development. They also promote long-term advancement of the field of humanitarian GIS.

The tutorial can be used as a primary reference for hands-on training in humanitarian GIS and as a supplemental reference for theoretical courses in humanitarian affairs, geography, and emergency management.

The following course scenarios are considered in the tables below:

- 5 weeks, 9 contact hours per week (intensive short-course schedule)
- 10 weeks, 4.5 contact hours per week (normal semester-course schedule)

GIS Tutorial for Humanitarian Assistance Short-course schedule 5 weeks, 9 contact hours per week			
Date	**Chapter**	**Theme**	**Key Learning Objectives**
Week 1	Chapter 1	Introducing humanitarian GIS applications	Understand potential for humanitarian GIS and review basic ArcMap navigation and data layer manipulations.
	Chapter 2	Mapping thematic data to support humanitarian operations	Build core cartographic skills, including data representation, layout design, and symbology.
Week 2	Chapter 3	Developing essential skills in spatial data management	Build core skills in data collection, preparation, storage, and sharing.
	Chapter 4	Generating spatial data during humanitarian emergencies	Discover the range of data sources for populating a geodatabase.
Week 3	Chapter 5	Using data models to create your own geodatabase	Encourage disciplined, standardized data management.
	Chapter 6	Managing hazardous operations using spatial data transformations	Build core skills in proximity and overlay analysis.
Week 4	Chapter 7	Planning logistics using network analysis	Demonstrate route optimization and catchment analysis.
Week 5	Chapter 8	Selecting sites with multiple spatial criteria	Advance skills in raster analysis, location optimization using multiple criteria, and automation of geoprocesses.
	Chapter 9	Improving the design and operation of refugee/IDP camps	Explore the potential of geostatistics.

GIS Tutorial for Humanitarian Assistance Semester-course schedule 10 weeks, 4.5 contact hours per week			
Date	**Chapter**	**Theme**	**Key Learning Objectives**
Week 1	Chapter 1	Introducing humanitarian GIS applications	Understand potential for humanitarian GIS and review basic ArcMap navigation and data layer manipulations.
Week 2	Chapter 2	Mapping thematic data to support humanitarian operations	Build core cartographic skills, including data representation, layout design, and symbology.
Week 3	Chapter 3	Developing essential skills in spatial data management	Build core skills in data collection, preparation, storage, and sharing.
Week 4	Chapter 4	Generating spatial data during humanitarian emergencies	Discover the range of data sources for populating a geodatabase.
Week 5	Chapter 5	Using data models to create your own geodatabase	Encourage disciplined, standardized data management.
Due date for midterm assignment: file geodatabase for term projects			
Week 6	Chapter 6	Managing hazardous operations using spatial data transformations	Build core skills in proximity and overlay analysis.
Week 7	Chapter 7	Planning logistics using network analysis	Demonstrate route optimization and catchment analysis.
Week 8	Chapter 8	Selecting sites with multiple spatial criteria	Advance skills in raster analysis, location optimization using multiple criteria, and automation of geoprocesses.
Week 9	Chapter 9	Improving the design and operation of refugee/IDP camps	Explore the potential of geostatistics.
Week 10	Class presentations of student projects		

Assignments and term project

Each chapter of GIS *Tutorial for Humanitarian Assistance* provides an opportunity for the student to demonstrate mastery of that chapter's content. These assignments should be a part of both the short-course and semester-course schedules above; however, it may be necessary to reduce the number of mandatory assignments in accelerated or large-group education settings.

A term project, assigned at the beginning of the course and designed to enable students to apply their knowledge using their topic of interest, is especially recommended for semester-long training.

Instructor resources DVD

A DVD containing a set of resources designed for instructors that are using, or planning to use, *GIS Tutorial for Humanitarian Assistance* in the classroom is available from the author. Resources on this DVD include lecture slides, lesson plans, and additional readings for each of the nine chapters. Sample results from actual students and a complete Blackboard-enabled curriculum, are also available.

If you would like to obtain this DVD, please send an e-mail to fverjee@gwu.edu with the subject line "GISHUM Teacher Resource Disk." Include your name, title, affiliation, phone number, and shipping address.

Acknowledgments

Producing this work was a labor of love for nearly three years. I wish to thank all those who contributed to its completion and who share my conviction that the application of GIS can significantly improve the impact of humanitarian operations.

First, I wish to recognize the team at ESRI Press. When I first proposed my concept to them, Peter Adams and his staff quickly recognized the need to capture the doctrine of humanitarian GIS. I am grateful to everyone at ESRI who made that possible, including my editors, technical reviewers, producers, and marketing analysts.

My sincere appreciation goes to Thomas Mazzuchi, chair of the Engineering Management and Systems Engineering Department; and Greg Shaw, director of the Institute for Crisis, Disaster, and Risk Management, at The George Washington University. Their encouragement and support were essential to the completion of this publication, and I am truly indebted. My special thanks also go to the students and peer reviewers who tested earlier drafts and who shared their candid opinion of what worked . . . and what didn't!

Most crucial to ensuring the realism of the case studies used were the many contributors of spatial data and operational documents. Obtaining viable spatial data from actual humanitarian emergencies was my single biggest challenge in this project. But I was reluctant to employ hypothetical scenarios lacking the richness and emotion of real humanitarian operations. I therefore extend my gratitude to Michael Sheinkman and his colleagues at the World Food Programme; Nigel Woof and Naomi Morris of MapAction UK; Shawn Messick and Olivier Cottray of iMMAP; and Yann Rébois and Bernard Wright of CartONG. They generously shared their expertise with me, most formidably Olivier, who contributed the draft manuscripts for chapters 5 and 7. They also patiently endured the lengthy period required to produce this work, and for that I am especially thankful.

Finally, I extend deep appreciation to my production assistant, Nuala Cowan of The George Washington University's Department of Geography. I could not have asked for a more capable interlocutor to help me realize my vision of this project, nor a more competent technician to help me hone the various exercises herein. Nuala was always resolute in her enthusiasm and tireless in her willingness to make my burden lighter.

We all hope that this book makes a difference.

OBJECTIVES

Understand cartographic and analytical applications of GIS
Manipulate map layers
Create spatial bookmarks
Use magnifier and overview windows
Measure distances
Identify features
View attribute tables
Load, unload, and create data layers
Make an orientation map

Chapter 1

Humanitarian GIS: Using basic tools and concepts

This first chapter familiarizes you with some of the basic mapping functionality of ArcGIS software in preparation for the more advanced chapters that follow. After a brief introduction to the various humanitarian applications of GIS, you will use the ArcMap application to create an orientation map that shows the location of an earthquake that occurred in Pakistan in 2008.

You may wish to repeat the exercises several times to ensure that you are comfortable with navigating within ArcMap and using its basic tools before progressing further.

What is GIS?

A geographic information system (GIS) is generally defined as the software, hardware, procedures, data, personnel, and interconnecting network that facilitate the input, storage, processing, analysis, and presentation of spatial data. This definition, which emphasizes the role of network-centric GIS, reflects how important Internet-based services have become in recent years.

What is spatial data?

All information that is associated with geographic values can be considered to be spatial data. A satellite image of the earth is an example of spatial data stored in raster format. A line representing roads is an example of spatial data stored in vector format. Vector data—points, lines, or polygons—enables attribute information to be easily associated with geographic features. For example, a line representing the road can be linked to an attribute table containing fields that describe road properties such as the number of lanes, seasonal serviceability, and surface type. Similarly, a set of points representing villages could be linked to an attribute table that describes each village's name, district, population, and postal code. Spatial databases are used to manage raster and vector data, and to connect that data with nonspatial data (e.g., textual reports, tabular files, etc.) that is not georeferenced.

Spatial data comes in three basic forms:

Map data: Map data contains the location and shape of geographic features. Maps use three basic shapes to present real-world features: points, lines, and areas (called polygons).

Attribute data: Attribute (tabular) data is the descriptive data that GIS links to map features. Attribute data might contain village names and district populations, and often comes packaged with map data. When implementing a GIS, the most common sources of attribute data are your own organization's databases combined with datasets you buy or acquire from other sources to fill in gaps.

Image data: Image data ranges from satellite images and aerial photographs to scanned topographic maps (maps that have been converted from printed to digital format).

ArcGIS Desktop software

ArcGIS Desktop is a suite of GIS products available in ArcView, ArcEditor, and ArcInfo license levels. The different licenses look and work similarly but vary in how much they can do.

ArcView, the most widely used GIS license in the world, is a full-featured GIS software application for visualizing, managing, creating, and analyzing geographic data. ArcEditor adds more GIS editing tools to ArcView. ArcInfo is the most comprehensive GIS license, adding advanced data conversion and geoprocessing capabilities to ArcEditor. While some of the exercises in this book require ArcEditor (provided on the accompanying disk), most can be completed using ArcView.

ArcGIS Desktop offers numerous extensions that can be added to ArcView, ArcEditor, or ArcInfo, three of which—Spatial Analyst, Network Analyst, and Geostatistical Analyst—are used in this book.

ArcGIS Desktop consists of two application programs: ArcCatalog and ArcMap. ArcCatalog is a utility program that has file browsing, data importing and converting, and file maintenance functions (such as create, copy, and delete)—all with special features for working with GIS source data. For managing GIS source data, you will use ArcCatalog instead of the Microsoft Windows utilities My Computer and Windows Explorer.

GIS analysts use ArcMap to compose thematic and orientation maps to carry out a wide range of analysis to support decision makers in the field as well as at the headquarters level. A map composition is saved in a map document file with a name chosen by the user including the .mxd file extension. For example, in this chapter you will open GISHUM_C1E1.mxd, a map document already created for you.

A map document stores pointers (paths) to map layers, data tables, and other data sources for use in a map composition but does not store a copy of any data source. Consequently, map layers can be stored anywhere on your computer, local area network, or even an Internet server, and be part of your map document. In this tutorial, you will use data sources available from the data DVD accompanying this book, as well as data prepared by you from datasets downloaded from various Web portals. You can store both types of data on your computer's hard drive for use in the exercises.

GIS in humanitarian assistance

Now that we have a basic description of GIS, we can explore its role and its potential in humanitarian assistance. Considering GIS from a functional point of view, GIS enables humanitarian organizations to do the following:

- Visually understand and communicate what is happening
- Store and share various types of information efficiently
- Analyze information to reveal hidden spatial relationships that can have a decisive impact on the outcome of humanitarian operations

Visualization and communication

Today, the role of GIS is primarily to help decision makers visually understand the field of operations, in the form of orientation and thematic maps. Mapmaking (cartography) is an essential application of GIS in almost any humanitarian intervention. Maps may be used to communicate administrative boundaries, terrain, navigational routes, hazards, relief operations, and countless other types of information critical to humanitarian organizations. These maps are often distributed through the UN's Humanitarian Information Centers (HICs) and information-sharing portals on the Internet. Maps do not need to be static poster displays. The recent advent of interactive Web-based geovisualization tools, such as ArcGIS Explorer and Google Earth, now enables decision makers to understand the field of humanitarian operations dynamically.

Storage and data exchange

GIS also plays an important role as a storage and data-sharing mechanism. Because any type of humanitarian information can be associated with a location, a GIS can be most efficient in capturing, managing, and sharing that information. By consistently referencing all data with a geolocation (latitude, longitude, and sometimes elevation), a GIS can store and share very different types of data efficiently, as well as analyze relationships between datasets that might otherwise be incompatible. For example, an atlas could describe a region's earthquake vulnerability and include a list of disaster risk-reduction projects there quite easily, but can it calculate how well those projects are addressing the region's seismic risk? Within a GIS, such dynamic data comparisons can be made.

Spatial analysis

Spatial analysis is the process of identifying meaningful patterns in spatial data and drawing conclusions from them. The techniques may be fairly simple, such as measuring the distance between a refugee camp and a reliable source of drinking water. Or they may be highly sophisticated, such as modeling the impact of disaster scenarios faced by vulnerable mountain communities in order to plan preventive and mitigative strategies. Some basic categories of spatial analysis are listed below along with examples of the types of questions that can be answered.

Queries and measurements

- How many people live within 100 meters of a river's floodplain?
- What is the total straight-line distance between an airport and emergency relief drop zones?
- What percentage of the people located above an elevation of 3,000 meters have not yet received winterized tents?
- Are my refugee camp operations compliant with the UNHCR emergency handbook (or Sphere Humanitarian Charter [http://www.sphereproject.org]) guidelines for providing water and sanitation?

Transformations

- Where is the best location for a health clinic if it needs to be within 10 kilometers of surrounding villages, within 50 meters of a main road, and more than 500 meters from areas known to contain land mines?
- Which villages need to receive additional humanitarian relief, given their distance from roads, humanitarian hubs, and a major earthquake?

Optimizations

- What is the shortest route between my headquarters and each of our local project offices?
- What would be the best location for a maternity clinic if it needs to be within a 20-minute commute from surrounding villages by ambulance?

Geostatistics

- What is the best village in which to rent an office based on market rates and the weighted distribution of surrounding populations?
- Is there a relationship between a rise in cholera infection and the available sources of drinking water?

Hypothesis testing and simulation

- What will happen if the river flood level rises 2 meters in the next week?
- What percentage of local buildings will be seriously damaged or destroyed after an earthquake measuring 7 on the Richter scale?
- What will happen if a glacier lake outburst flood occurs in this valley?
- How are criminals likely to respond if we change our regimen of nightly security patrols?

Not all of these categories of spatial analysis are feasible or even necessary during most humanitarian interventions, but they do reflect the enormous range of ways in which GIS can improve decision making at the headquarters level and in the field.

Scenario: 2008 earthquake in Pakistan

In the following series of exercises, you will produce a map for a region that has just experienced a major earthquake and is likely to require international humanitarian assistance. Humanitarian emergencies often occur in remote locations unfamiliar to responders and decision makers. The purpose of the map, therefore, will be to illustrate the location of the disaster and its proximity to population centers and transportation infrastructure. You will accomplish the following:

- Display the location of the earthquake using latitude and longitude information from an external source
- Create and export an orientation map to a presentation for briefing officials

A data dictionary appears in each chapter in this book. **Develop the habit of studying the data dictionary before beginning the exercises.** Likewise, whenever you receive a new dataset in your real work, always start by studying that dataset's origins and contents before using it. Only then can you use it effectively in your application or share your outputs confidently with decision makers.

Table 1.1, the data dictionary for chapter 1, lists the relevant attributes of each data layer used in this chapter.

Layer[a] or attribute	Description
Major_Cities.shp	Pakistan major cities (points)
CITY_NAME	City name
ADMIN_NAME	Administrative region name
STATUS	National/provincial capital or not a capital
POP_CLASS	Population range classification
Populated_Places.shp	Pakistan gazetteer populated places (points)
NAME	Populated place-name
Admin_Boundaries.shp	Pakistan second administrative layer boundaries (polygons)
ADMIN_NAME	Administrative unit name
POP_ADMIN	Population
SQKM	Area in square kilometers
Earthsat_150m_Pakistan.jp2	NaturalVue Landsat mosaic for Pakistan and surrounding region
[a]Files containing the .shp extension are shapefiles. The last layer is a raster dataset.	

Table 1.1 Data dictionary

Exercise 1.1

Manipulating map layers

In this first exercise, you will learn how to open and move around an existing map document. Navigating within ArcMap and customizing the display of your spatial data are essential skills that don't take long to master. Try not to rush through these exercises if you are new to ArcGIS: take your time, experiment with the various tools introduced here, and develop some deftness before proceeding to the next exercise.

Open an existing map document

1. To launch ArcMap, from the Windows taskbar, click Start > All Programs > ArcGIS > ArcMap. (Alternatively, you can start ArcMap by double-clicking the ArcMap icon 🐢 on your computer desktop.)

2. If a welcome screen appears, select "An existing map" and press OK. If a welcome screen does not appear, click File > Open.

3. Browse to the location of the GISHUM folder (e.g., C:\ESRIPress\GISHUM), double-click the Chapter1 folder, and then click the GISHUM_C1E1.mxd document to select it.

4. Click Open.

Note: You can also open a map document by navigating to the file using Windows Explorer and then double-clicking the *.mxd file to launch ArcMap and load the associated layers.

The GISHUM_C1E1.mxd file contains four layers of data for Pakistan: two point layers showing the major cities and other populated places, one polygon layer showing the administrative unit boundaries, and one raster layer containing a satellite image of the region.

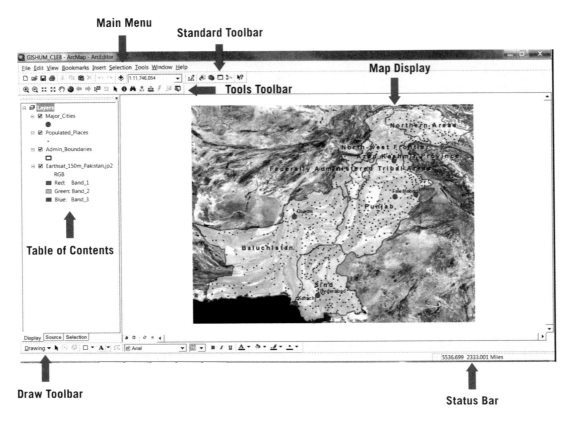

Before proceeding, let's review the major components of the ArcMap application window:

- The main menu contains the familiar drop-down arrangement common to all Windows applications; it provides access to ArcMap's functionality.
- The Standard toolbar is used for map printing, creating a new map, opening an existing map, and starting related ArcGIS applications.
- The Tools toolbar contains frequently used mapping tools to change the map extent, select features, measure distances, and find locations.
- The map display shows the map with all selected data layers visible.
- The Draw toolbar is used for annotating your map display with graphics and text.
- The Status bar shows the map coordinates of the cursor's location in the map display.

All toolbars can be undocked and dragged by their gray, vertical "handle" to any location on your screen and repositioned according to your preferences. If you cannot find a toolbar, click View on the main menu, and then click Toolbars.

View map layers

A digital map is composed of layers. You can see the layers listed in the table of contents on the left-hand side of the ArcMap window. This table of contents contains four layers. Notice the check boxes next to the layer names in the table of contents. When the box is checked, the layer is visible in the map. You can turn any of the layers on and off in the table of contents by clicking their check boxes.

1. **Click the minus icons to the left of the check boxes and notice how the table of contents changes (it works in a similar way to a data tree within Windows Explorer). Click the check box next to the Admin_Boundaries layer to turn the layer off; click it again to make it reappear.**

2. **Hold down your left mouse button over Major_Cities, and then drag it below the Admin_Boundaries layer.**

What happened to your map? The table of contents controls the various layers of your map. Whichever layer is on top in the table of contents will also be the topmost layer of your map.

3. **Return the Major_Cities layer to its position above the Populated_Places layer.**

Find toolbars

ArcMap organizes individual tools into related sets called "toolbars," which can be moved to different parts of the screen (floating) or docked in a specific location, in accordance with the user's preferences. If you move your mouse over any icon without clicking (often called "mousing over" or "hovering"), a pop-up box will show the name or function of that particular tool.

1. **Mouse over each of the buttons in your Standard, Tools, and Draw toolbars and familiarize yourself with some of the core functionalities provided by these toolbars.**

Locate Fixed-zoom tools

ArcMap has a suite of zoom and pan tools available on the Tools toolbar that allow you to view a map from a range of scales. These tools enable you to concentrate on a particular area of a map, reduce the confusion caused by overlapping features and labels in too small a scale, and then return to the full extent of your map to gain perspective.

1. **On the Tools toolbar, locate the Fixed Zoom In and Fixed Zoom Out tools.**

2. **Click the Fixed Zoom In ⅗ tool three times.**

Depending on the speed of your computer, it may take a moment for the layers to redraw at the larger scale. With each click the display zooms in on the center of the map in fixed scale increments.

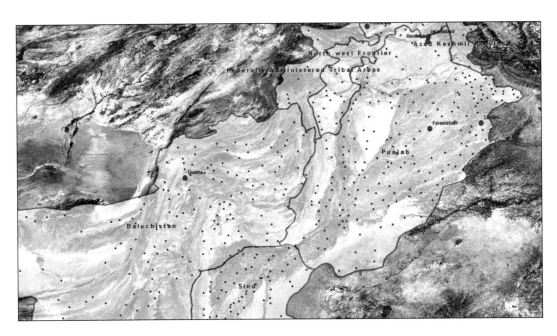

3. Click the Fixed Zoom Out tool three times to return the map to the previous scale.

Use custom zoom tools

1. On the Tools toolbar, click the Zoom In tool.

2. Click and hold the mouse button on a point above and to the left of the northwest tip of Baluchistan.

3. Drag the mouse to draw a box that extends to the southeastern corner of Sind.

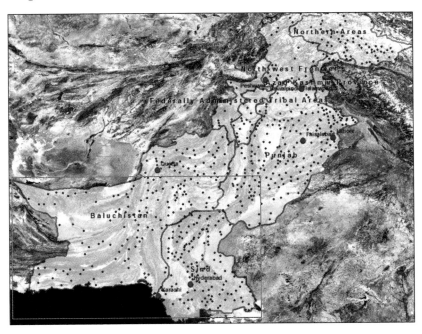

The resultant map is zoomed in on the southern portion of Pakistan. The major cities, other populated places, and the geology of the region are easier to discern.

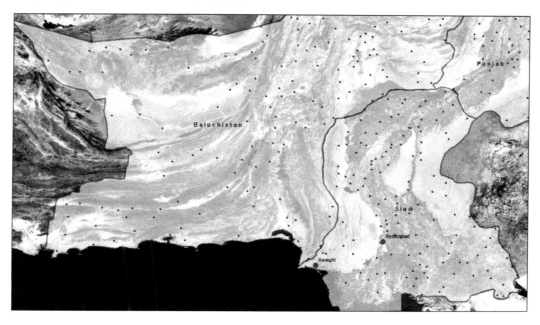

The Zoom In tool can also be used to click a specific point on the map.

4. **Using the Zoom In tool, click the Hyderabad symbol.**

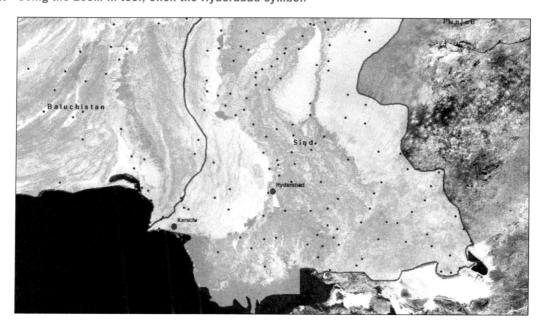

The resultant map is zoomed in and centered on the city of Hyderabad.

Use the scale box

On your Standard toolbar, you will see a scale box that provides you the option of manually typing in your map display's scale or selecting it from the drop-down menu.

1.   **On the Standard toolbar, click the drop-down button, and then select a scale of 1:1,000,000.**

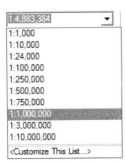

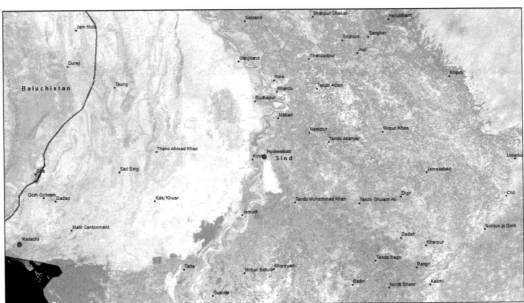

The resultant map is centered on Hyderabad at a 1:1,000,000 scale. In addition, the names for the Populated_ Places layer now appear. ArcMap provides the ability to set scale ranges that determine whether certain layers, and their associated labels, will be displayed.

Zoom to layer

ArcMap allows you to zoom to the full extent of any layer listed in the table of contents.

1. **In the table of contents, right-click the Major_Cities layer, and then click Zoom To Layer.**

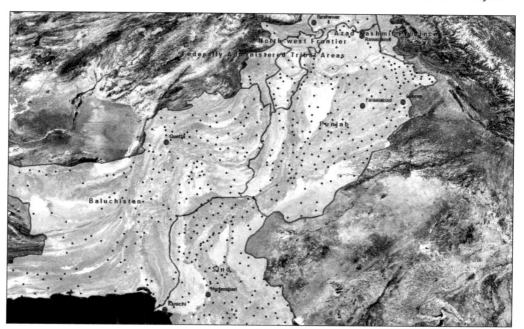

This time the map is redrawn to the smallest scale that shows all features contained in the Major_Cities layer. Since that scale is smaller than the visibility threshold of place-names associated with the Populated_Places layer, the latter's labels are turned off to improve the map's visibility.

Go back and forth between extents viewed

As you change your map display, ArcMap records each extent setting and allows you to go back and forth between the extents you have viewed.

1. **On the Tools toolbar, click Go Back To Previous Extent ⬅ and Go To Next Extent ➡ .**

2. **Click these buttons to move back and forth between the map extents you have viewed.**

Use the Full Extent tool

The best way to ensure that you are looking at all of the data in your map display is to use the Full Extent tool.

1. **On the Tools toolbar, click the Full Extent 🌐 tool.**

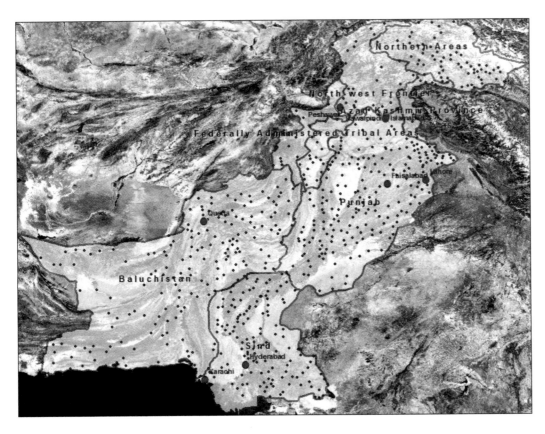

The resultant map zooms out to contain every layer that is displayed. The Full Extent tool is especially handy if you cannot find some data that has been projected to a region outside of where you expected to see it.

Zoom using the mouse wheel

Using the mouse wheel to zoom in and out of the map display provides a quick and easy way to focus on a particular location.

1. **Mouse over any location of interest in your map display and, without clicking, roll the mouse wheel towards you to zoom in to that part of the map.**

2. **Roll the mouse wheel away from you to zoom out from that part of the map.**

Pan

You can also move around the map without changing the scale by using the Pan tool.

1. **On the Tools toolbar, click the Pan tool.**

2. **Click the middle of the map display and, pressing on the mouse button, drag the map to the bottom edge of the display.**

3. **Release the mouse button.**

The map display has shifted northward, though its scale remains the same. The Pan tool can be used to move in any direction within the current data frame without changing the scale.

Create spatial bookmarks

A spatial bookmark identifies a particular geographic location that you want to save and reference later. For example, you might create a spatial bookmark that identifies a study area. As you pan and zoom around your map, you can easily return to the study area by accessing the bookmark. You can also use spatial bookmarks to highlight areas on your map you want others to see.

1. **Zoom to the full extent of the data frame.**

2. **Zoom in on the city of Karachi to a scale of 1:250,000.**

3. **On the main menu, select the Bookmarks menu.**

4. **Click Create, and then name the bookmark Karachi. Click OK.**

5. **Zoom to the full extent of the data frame.**

6. **Click the Bookmarks menu, and then select Karachi.**

ArcMap zooms to the exact extent of the saved bookmark for Karachi. This bookmark is now available every time you use this map document, which is helpful if you frequently zoom to this extent.

Exercise 1.2

Using the Magnifier and Overview windows

Sometimes it is helpful to have both a larger and smaller scale display of your region of interest in order to navigate effectively. In this exercise you will learn how to use the Magnifier and Overview tools provided in ArcMap.

Use the Magnifier window

The Magnifier window works like a magnifying glass: as you pass the window over the data, you see a magnified view of the location under the window. Moving the window does not affect the current map display.

1. **Starting from your Karachi bookmark extent, use the scale box to zoom to a scale of 1:1,000,000.**

2. **On the main menu, click Window and then Magnifier.**

The Magnifier window appears at an unspecified location over the map display at the default percent magnification level.

3. **Click the title bar at the top of the Magnifier window, and then drag it over the map display until the crosshairs are positioned over Naka Kharari.**

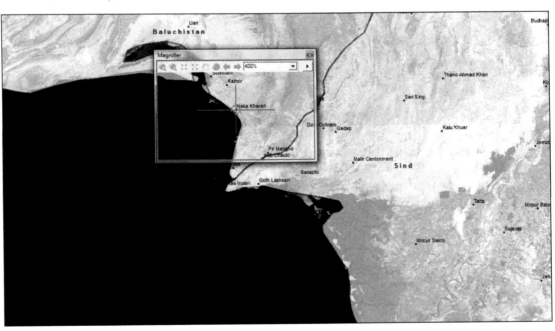

4. Release the mouse button to see the zoomed detail of the map at this location.

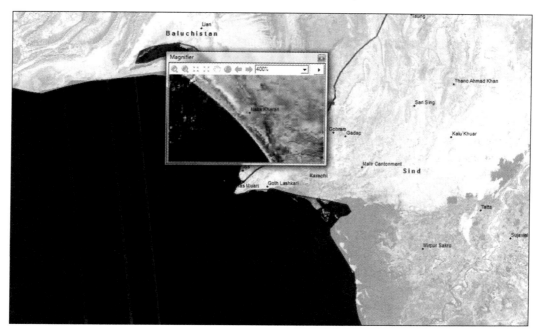

5. Click the drop-down arrow, and then increase the magnification to 600 percent.

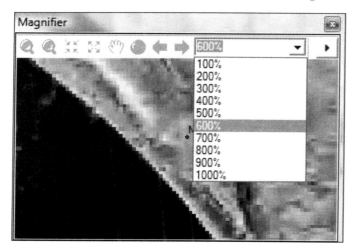

The magnifier zooms farther into the selected position on the map display. You can also see the magnification dynamically as you pan the window over the map.

6. Click the arrow in the upper right of the Magnifier window.

7. Select Update While Dragging.

8. Drag the window around the map display to see the magnification as it updates.

9. Close the Magnifier window.

Use the Layers Overview window

The Layers Overview window provides a thumbnail view of the full extent of the map (for all layers). A red box in the Layers Overview window represents the currently displayed area on the map. By left-clicking and holding with your mouse, you can move this red box to pan the map display and make the box smaller or larger to zoom the map display in or out.

1. **Zoom to the full extent of the data frame.**

2. **Zoom to the city of Quetta at a scale of 1:3,000,000.**

3. **On the main menu, click Window and then Overview.**

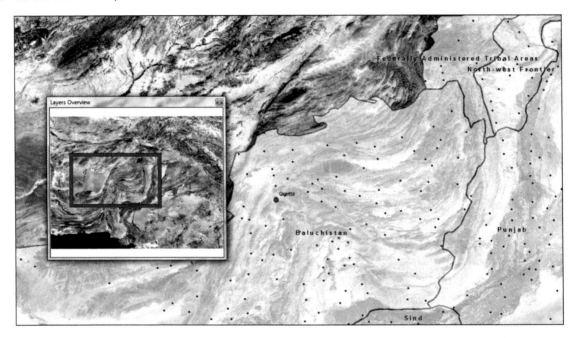

The current extent of the map display is highlighted in the Layers Overview window with a red box.

4. **Position the cursor at the center of the red box, and then click and drag to move the map display toward India (southeast limits of the display).**

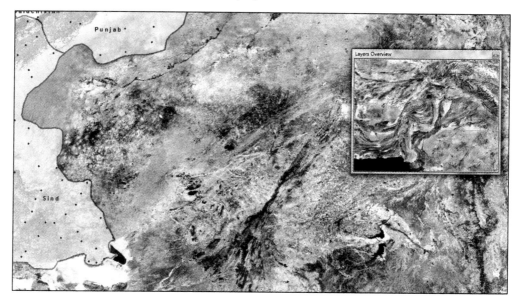

The extent of the map display updates to reflect the changes made as you moved the Layers Overview window to a new location.

You can set options for the Overview window by right-clicking its title bar and clicking Properties. For example, you can choose the box's fill symbol and the window's background color. In addition, you can choose which layer is shown in the Overview window. By default, the layer at the bottom of the ArcMap table of contents is the one drawn in the Overview window and is used to set the extent of the Overview window. You can modify the reference layer's properties from the Overview Properties dialog box; any changes you make in the Layers Overview window are also applied in the map and vice versa.

5. **Close the Layers Overview window.**

Exercise 1.3

Measuring distance

The Measure tool allows you to measure the distance of a line or the area of a polygon. This tool is very helpful when estimates of ground distances are needed for humanitarian planners responsible for the logistics and staging of an emergency response.

Use the Measure tool

1. **Zoom to the full extent of the data frame.**

2. **Zoom in on the region of Pakistan containing the cities of Islamabad and Lahore at a scale of 1:3,000,000.**

3. **On the Tools toolbar, click the Measure** 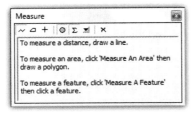 **button.**

The Measure window appears, with instructions on measuring the distance of a line, the area of polygon, or a feature. It defaults to the Line Measurement tool.

4. **Click the symbol for Islamabad, and then click the symbol for Lahore.**

The Measure window indicates that the straight-line distance is approximately 2.5 decimal degrees. (If your default units of measurement are not decimal degrees, you will get a different value stated in meters, miles, etc.)

5. **Now click the symbol for Faisalabad.**

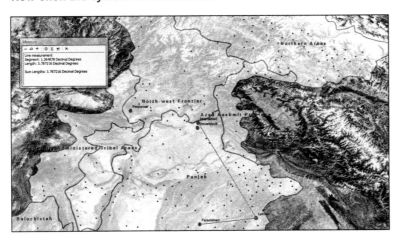

The measurements in the window show the length of the last segment and the sum of the lengths of all the segments in decimal degrees, or whatever default units of measurement you are using. This unit of measure is used because the map is set in geographic coordinates.

Change measurement units

You can change the units to a more appropriate measure.

1. **Double-click the Faisalabad symbol to complete the segment being measured.**

2. **Click the Choose Units** ![button] **button on the Measure menu.**

3. **Select Distance and then Kilometers.**

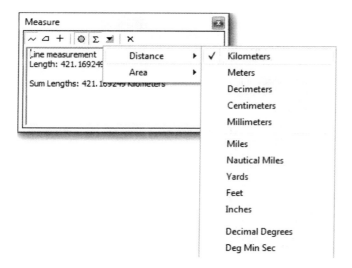

The measurements are now displayed in kilometers, which are more useful than decimal degrees for planning purposes.

4. **Close the Measure box.**

To measure the area of a polygon or feature, the data must be in a projected coordinate system. The four layers in this chapter are not projected. In chapter 3 you will learn how to apply various geographic and projected coordinate systems.

Your turn

Use the Measure tool to determine distances between other features in the map in various units of measurement.

Exercise 1.4

Identifying features

The Identify tool displays the data attributes of any feature, or group of similar features, clicked in the map display.

Identify a single feature

1. **Zoom to the full extent of the data frame.**

2. **On the Tools toolbar, click the Identify 🛈 tool.**

3. **Click the symbol for Quetta.**

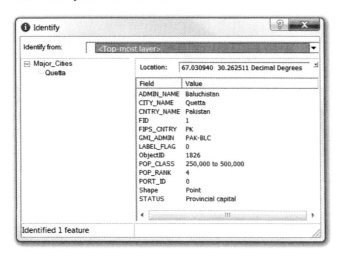

The Identify window populates with the attribute information associated with Quetta. Be sure that the <Top-most layer>, or Major_Cities, is the layer to identify from.

The layer tree shows that Quetta is the provincial capital of Baluchistan, is located at approximately 67.030940 east longitude and 30.262511 north latitude in decimal degrees (coordinates of longitude and latitude, your values will be slightly different), and had a population of 250,000 to 500,000 when the attribute data was collected.

Identify multiple features

You can also view the attributes of multiple features at once by including them in a bounding box drawn with the Identify tool.

1. **Drag the Identify tool over the symbols for Islamabad, Rawalpindi, and Peshawar.**

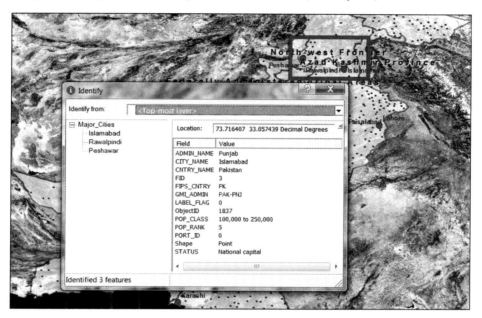

2. **Click the three city names in the layer tree to flash the location and view the attributes of each city.**

Control results obtained with the Identify tool

So far you have only viewed the attributes of features within a single layer. The Identify tool defaults to the topmost layer in your table of contents, but this setting can be modified to select another layer or multiple layers.

1. **Select <All layers> from the "Identify from" drop-down list.**

2. **Use the Identify tool to click the symbol for Peshawar.**

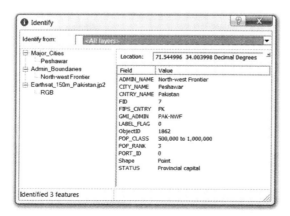

The Identify window is populated with attribute data for all of the layers that intersect at the symbol for Peshawar.

3. Click each of the three layers in the data tree to view attribute information.

4. Click the Field column heading to sort the fields in ascending or descending order. Do the same to sort the entries in the Value column.

5. Close the Identify box.

Your turn

Use the Identify tool to view the attributes of other single and multiple features in your map. Then view the attributes of one, some, or all visible layers at a certain location in the map.

Exercise 1.5

Viewing attribute tables

All geographic features in a GIS have attribute tables associated with them. The data in the attribute tables determines the geometry of the feature (point, line, or polygon), the location (i.e., point or centroid coordinates), and other descriptive information. Attribute tables contain a record for every feature in the layer (shown in rows) and for every field of information (shown in columns).

When you clicked the cities of Quetta and Peshawar in the previous exercise, the Identify tool searched the attributes of the Major_Cities layer (and underlying layers) to produce the results you saw.

Open an attribute table

1. **In the table of contents, right-click the Admin_Boundaries layer.**

2. **Select Open Attribute Table.**

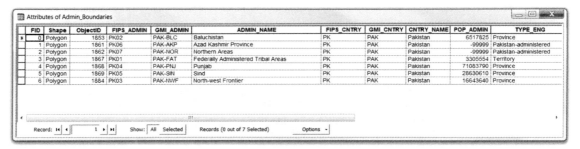

The table has seven records, one for each major administrative region in Pakistan, and 16 fields representing different categories of data about each region (you may need to scroll to the right of your window or increase its width in order to see all the columns). The FID field contains a unique identification number for each record, or administrative unit. The Shape field describes the feature's geometry (point, line, or polygon). Among other attributes are the name, type, population, and area (in square kilometers and in square miles) for each administrative unit.

Move a field in an attribute table

Sometimes you may need to reposition fields in an attribute table to make them easier to view. For example, you may want to directly compare the populations and areas of all provinces listed in the table.

1. **Click the POP_ADMIN column heading in the attribute table.**

2. **Hold down the left button on the mouse, and drag this column to the right side of the ADMIN_NAME column.**

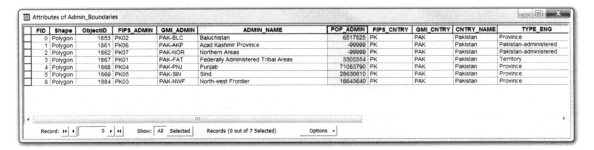

3. **Now drag the SQKM column to the right side of the POP_ADMIN column.**

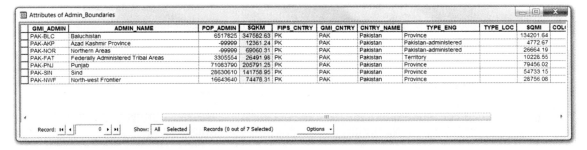

The population and area fields are now positioned next to the name of each administrative region. Rearranging the data like this has no adverse effects on the database or map.

Sort a field in an attribute table

In analyzing attribute data, it can be helpful to sort the fields in alphabetical or numerical order. For example, you may want to rank Pakistan's administrative regions on the basis of population.

1. **Right-click the POP_ADMIN column heading, and then select Sort Ascending.**

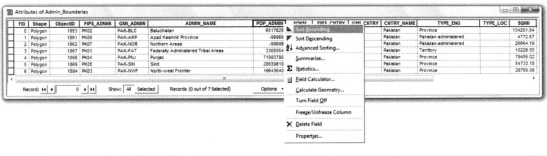

The administrative regions are now sorted in order of increasing population. The population records for Azad Kashmir and Northern Areas reflect the fact that these two regions are internationally disputed and accurate figures are unavailable.

Your turn

Sort the administrative regions of Pakistan by size in square kilometers in descending order. Then sort city names in the Major_Cities attribute table by alphabetical order.

Select a feature from the attribute table

Records, just like fields, can be selected. When a record is highlighted in a table, its corresponding feature is highlighted in the map. A highlighted record or feature is said to be "selected."

1. **Open the Populated_Places attribute table, and then sort the entries in alphabetical order by name.**

2. **Click the small gray tab at the far left edge of the second record in the table, Abbottabad.**

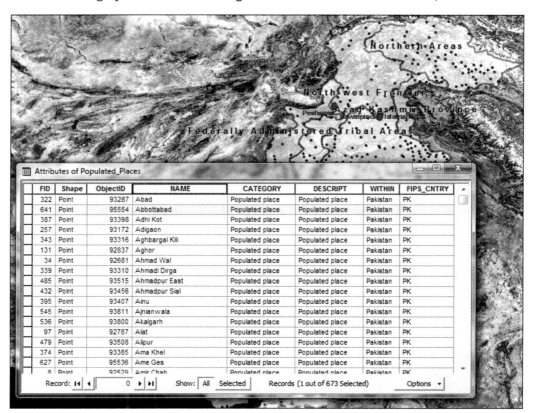

The selected record is now highlighted in blue. When one or more records are selected in a table, they will also be selected on the map, and vice versa.

3. **To cancel the selection, click the Options tab at the bottom of the table, and then select Clear Selection.**

Select a feature in the map display

You can also select objects in ArcMap directly from the map window.

1. Close the attribute table.

2. Zoom to full extent.

3. On the Tools toolbar, click the Select Features 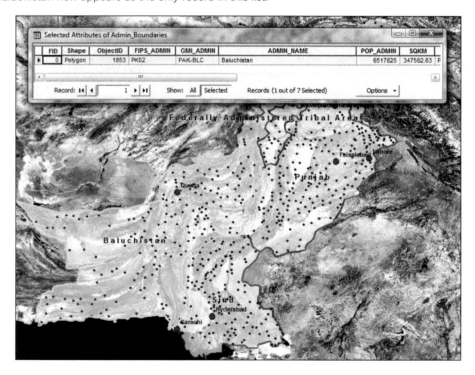 tool.

4. Select the large region in the southwest of the country.

5. Open the Admin_Boundaries attribute table.

6. At the bottom of the attribute table, click the button labeled Selected.

Baluchistan now appears as the only record in this list.

7. Cancel the selection using the Options tab. Close the attribute table.

8. Select several populated places in the map display by holding down the Shift key of your keyboard while you click each black dot.

9. Open the Populated_Places attribute table, and then show only the selected records.

10. Cancel the selection by clicking the Clear Selected Features ⊠ button on the Tools toolbar. Close the attribute table.

Exercise 1.6

Loading and unloading data layers

Sometimes you may need to remove and add layers to your map display using the following manual process.

Remove data from the table of contents

1. **Right-click the Populated_Places layer.**

2. **Select Remove.**

The layer has been unloaded from ArcMap. Because we need that layer for the remainder of this chapter, let's load it back into the table of contents.

Add data to the table of contents

1. **From the File menu, select Add Data (the Add Data button is replicated on the Standard toolbar, as is the case for many of the tools and functions in ArcMap).**

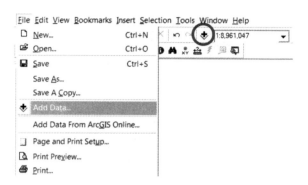

2. In the Add Data pop-up window, click the Connect To Folder button.

3. In the Connect To Folder window, navigate to the location of the book's folder (e.g., C:\ESRIPress\ GISHUM), and then click OK.

You will have to perform this step only once. By establishing a connection, you will be directed to this folder each time you wish to add new data to the map.

4. Open the Chapter1 folder, select the file Populated_Places.shp, and click Add.

The populated places layer (with a different, random default symbology) now appears at the top of your table of contents.

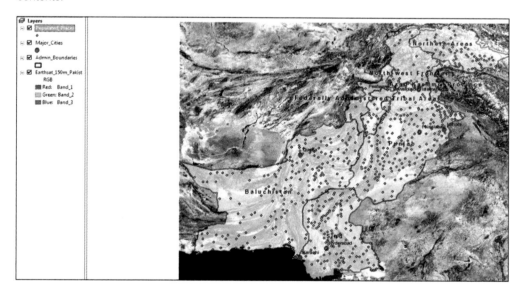

Your turn

Practice unloading and loading each of the layers in your Chapter1 folder. Note that larger data files, like the EarthSat imagery, take much longer to load than smaller data files, like the point feature layer representing Pakistan's major cities.

Exercise 1.7

Creating a new data layer

In most rapid-onset emergencies, the locations of disaster-stricken areas are usually communicated as place-names or geographical coordinates. For earthquakes, the U.S. Geological Survey (USGS) provides real-time and historical data in a variety of data formats for any location around the world (http://earthquake.usgs.gov/).

In this exercise, you will plot the location of a magnitude 6.4 earthquake that occurred near Quetta, Baluchistan, at 23:09:58 UTC on October 28, 2008. It, together with another earthquake the next day, killed 166 people and injured 370 more. Several villages were destroyed by landslides, 3,487 homes were destroyed, and 4,125 homes were damaged (USGS 2008).

You have just retrieved the earthquake epicenter's latitude and longitude coordinates (30.656°N, 67.361°E) from the USGS Web site and will now add that information to your map display.

Go to XY location

The first step is to plot the epicenter on your map. You can use the Go To XY command on the Tools toolbar to navigate to a particular x,y location in your map. You can specify the location by entering coordinates in decimal degrees, degrees-minutes-seconds, degrees-decimal-minutes, the Military Grid Reference System, or the U.S. National Grid System.

1. Open the map document **GISHUM_C1E8.mxd**. This will return you to your original map display settings. (Click Yes when asked if you wish to save changes to your initial map document.)

2. On the Tools toolbar, click the Go To XY $\overset{\oplus}{\text{xy}}$ button.

3. Click the Units button, and then view the various coordinate systems. Select Decimal Degrees, if it's not the default.

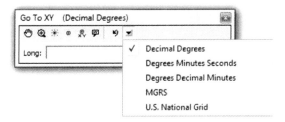

4. To find the earthquake's epicenter, enter the longitude (Long) **67.361** (x-coordinate) and latitude (Lat) **30.656** (y-coordinate).

5. Click the Zoom To button on the Go To XY toolbar to locate the region of the October 2008 earthquake.

6. Click the Flash button to identify the precise location once again.

7. **Click the Add Point [image] button to mark the earthquake's epicenter. If your map label for Baluchistan covers the epicenter symbol, zoom or pan your map display as required to reveal the symbol.**

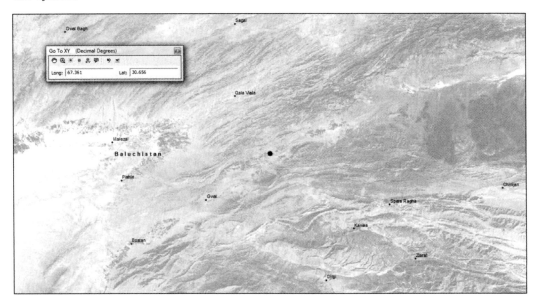

A point marker is added to the map display. Because the default marker blends into the rest of the map display, you will change its properties.

Change point marker symbol properties

1. **Using the Select Elements [image] tool on the Tools toolbar, click the earthquake point marker, and then delete it using the Delete key on your keyboard. We will re-create it using more effective symbology.**

2. **On the Draw toolbar, click the Drawing menu and then Default Symbol Properties.**

3. Click the Marker Properties button to open the Symbol Selector.

4. Scroll down and click the Circle 22 symbol.

5. Select the color Solar Yellow and size 25.

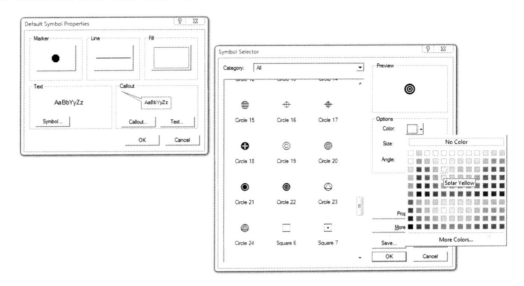

6. Click **OK** to close the Symbol Selector, and then click **OK** to close the Default Symbol Properties.

7. On the Go To XY toolbar, click the Add Point button again.

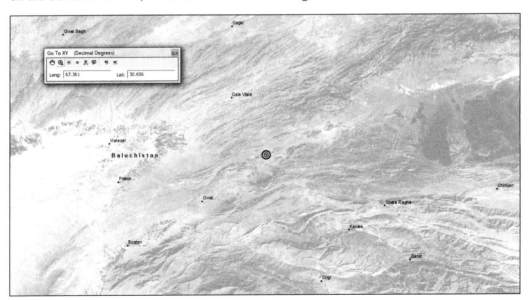

The epicenter is now more obvious. You will now convert it from a graphic into a feature layer.

Convert graphics to features

1. **Close the Go To XY toolbar.**

2. **On the Draw toolbar, click the Drawing menu, and then select Convert Graphics To Features.**

3. **In the Convert Graphics To Features dialog box, ensure that your conversion preferences are as follows:**

   - Convert: Point graphics
   - Use the same coordinate system as: the data frame
   - Output shape file or feature class: Earthquake.shp (to go into your Chapter1 folder)
   - Automatically delete graphics after conversion: Yes (checked)

4. **When your dialog box matches the image below, click OK.**

5. **Click OK when asked if you want to add the export data to the map as a layer.**

6. **Zoom to the full extent of the map display.**

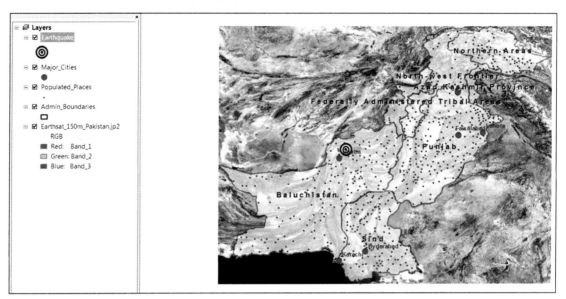

The map display now shows the earthquake's location within Pakistan, and the marker now appears as a point feature (vector) layer in your table of contents. The earthquake's location has been saved as a separate shapefile in your data folder.

Hyperlink photos or URLs to maps

ArcMap enables you to hyperlink images or Internet addresses to the map so that they are displayed when a user clicks an associated feature. This tool can be very useful during and after humanitarian operations for verifying the status of particular locations affected by an emergency.

1. **Zoom in toward the earthquake marker until you can see the village of Ziarat.**

2. **Using the Identify tool, click the symbol representing Ziarat.**

3. **Right-click Ziarat in the left window pane of the Identify window, and then select Add Hyperlink.**

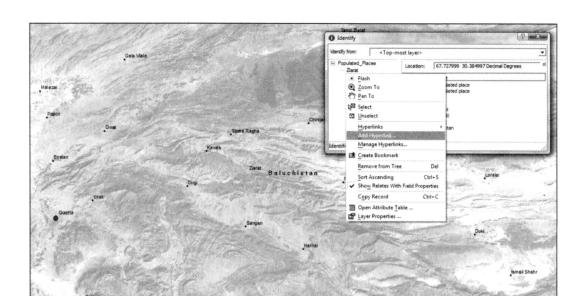

4. **Select Link to a URL, and then enter the following hyperlink:**

   http://www.irinnews.org/Report.aspx?ReportId=81622

5. **Click OK, then close the Identify window.**

6. **On the Tools toolbar, click the Hyperlink** ⚡ **tool.**

7. **Click the symbol representing Ziarat.**

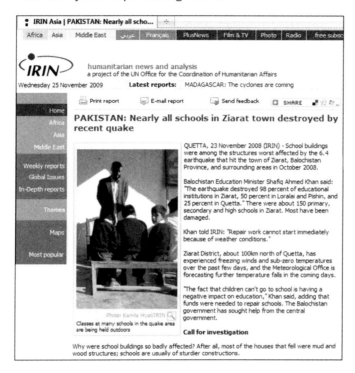

Your Internet browser will launch and navigate to the hyperlinked Web site. The hyperlink function is not limited to Web sites. Documents that are stored on your own computer, such as word-processing files, presentations, spreadsheets, and photographs, can also be linked to map features, making them more accessible in the future.

Exercise 1.8

Making an orientation map

You now have a simple orientation map showing the location of the October 2008 earthquake near Quetta, Pakistan. Your final step is to prepare the map for export or printing.

Change data view to layout view

The map window that you have been working in until now is called Data View. You will now switch to a different map window, called Layout View, to prepare the map aesthetics.

1. **Zoom to full extent.**

2. **Click the Layout View** ▯ **tab in the bottom left-hand corner of the main map window.**

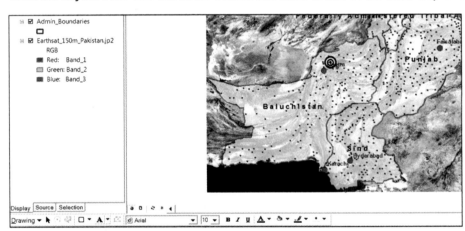

3. **Now click the Data View** 🌐 **tab (to the left of the Layout View tab). Observe the difference.**

4. **Return to the Layout View window.**

The Refresh View 🔁 tab (to the right of the Layout View tab) can be used to ensure that your map is up to date with respect to any recent changes made to it.

Apply a map template

The fastest method to generate a professional-looking map is to use one of the map templates included with ArcGIS or a custom template developed for your organization. Such templates typically contain a predefined page layout that arranges map elements such as north arrows, scale bars, and logos. All you need to do is prepare your map display and apply the template, and you are ready to export or print the map.

1. **To apply a new template to an existing layout, click the Change Layout** ⊞ **button on the Layout toolbar.**

2. Explore the various templates, and then choose the Asia.mxt template from the World category.

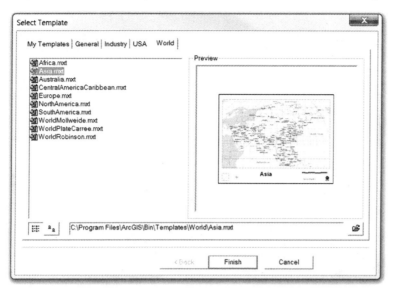

3. Click Finish.

4. If the top of your legend is hidden, right-click the portion under the map, and then select Order > Bring to Front.

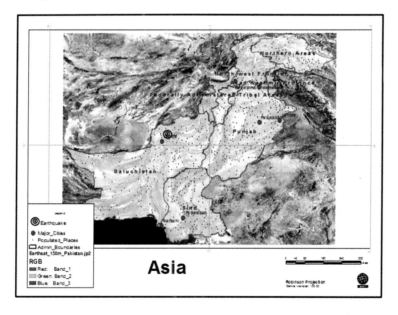

The template has been applied to your map and is ready for customization.

5. On the Standard toolbar, set the map scale to 1:10,000,000 (this will ensure that your map fills the available space in the layout).

6. Double-click the "Asia" title box and rename the map **Pakistan**.

7. Double-click the Legend box, and then click the Items tab.

8. Click the Earthsat_150m_Pakistan.jp2 layer under Legend Items, and then click the Remove button.

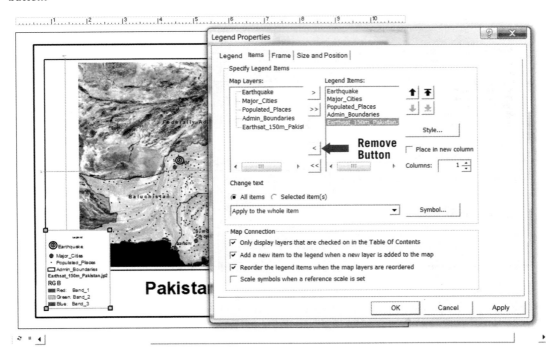

9. Click the Legend tab, and then clear the Show option.

10. Click OK to accept your new legend properties.

11. Left-click the data frame to select it.

12. Right-click the data frame, and then click Properties.

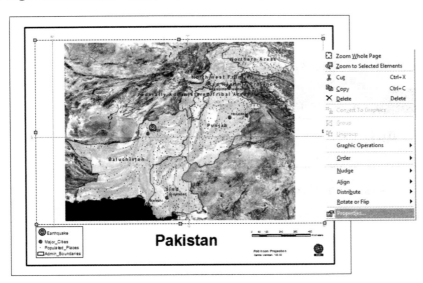

13. Click the Grids tab, and then click Properties.

14. On the Intervals tab, change the X and Y axis intervals from 10 degrees to 5 degrees.

15. Click OK.

16. In the Data Frame Properties dialog box, select the Data Frame tab, and then change the radio button to Fixed Scale.

17. Click OK to close the Data Frame Properties dialog box.

18. Double-click the scale bar.

19. On the Scale and Units tab, change Division Units to Kilometers. Click OK.

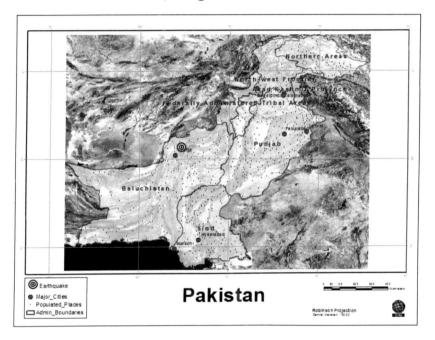

Your map is now ready to be exported for printing. As an alternative to applying a map template (as you did above), you can add elements and customize your map manually by clicking Insert on the main menu.

Export your map

It is often useful to export a finished map to a file format that preserves its appearance and makes it accessible to anyone regardless of his or her GIS skill level. When exporting from ArcMap to PDF, be sure to embed all fonts in the document.

1. On the main menu, click File, and then select Export Map.

2. Export your map to your Chapter1 data folder as **PakistanEQ1.pdf.**

Your turn

Practice changing and customizing your map layout settings. At a scale of 1:1,000,000, the region surrounding the epicenter appears much larger, and several changes become necessary: the grid spacing needs to be smaller, the map title needs to change, the Admin_Boundaries layer in the Legend box becomes redundant, and so forth. Regenerate your map to be effective at this finer scale and save it as **PakistanEQ2.pdf.**

You have completed your first exercise in ArcMap and successfully exported your map product.

What to turn in

If you are working in a classroom setting with an instructor, submit electronic copies or printouts of your map at scales of 1:10,000,000 and 1:1,000,000.

Assignment

Write a short essay (5–7 pages) that describes how an organization that provides humanitarian assistance uses GIS. Consider humanitarian agencies within the United Nations, a national emergency management agency, or a nongovernmental organization.

In your research, identify how your chosen organization applies GIS to support its mission, and explain the process it uses to generate GIS products and services. Also describe the size and expertise of its GIS program staff, its primary types and sources of GIS data, and how it delivers its products and services to end users. Finally, consider the organization's use of the Internet for finding GIS data, publishing finished products, or facilitating Web-based GIS.

What to turn in

If you are working in a classroom setting with an instructor, submit an electronic copy or printout of your essay.

References

USGS. 2008. Tectonic summary. http://earthquake.usgs.gov/earthquakes/eqinthenews/2008/US2008yscs/#summary (accessed October 8, 2008).

Chapter 2

Mapping thematic data

One of the most essential applications of GIS is to create a visual representation of complex data so decision makers can grasp the underlying themes quickly. Maps are an effective way to communicate and analyze information and to present thematic data intelligently.

Rendering data visually takes practice, creativity, and good cartographic technique, an essential skill for GIS professionals working in the field of humanitarian assistance. ArcMap offers unparalleled utility for the skilled cartographer, and with it you will begin to learn how to produce professional, informative maps in this chapter.

Thematic maps for humanitarian operations

A thematic map is a visual representation of one or more sets of data, or themes, illustrating their distribution and spatial relationships. Thematic maps usually focus on just one or a few themes, with only minimal reference information to provide the map reader with geographic context. (That is the purpose of orientation maps, like the one you produced in chapter 1.)

Examples of thematic maps include weather maps, population density maps, crisis maps, agricultural production maps, and more. Characteristics of thematic maps include the following:

- They display limited themes of information.
- The primary aim is communication of a specific topic.
- The graphic marks (that is, symbolized points, lines, or polygons; text; or raster pixels) are designed to draw the reader's attention to important aspects of the distribution being mapped.

The mapmaker uses carefully considered symbology and labeling to direct attention to the significance in the distribution of one or a few geographic phenomena. This task becomes substantially more difficult if more than one theme of information is included on a thematic map, as in a bivariate or multivariate map.

The emphasis of thematic maps is on the geographic pattern of the feature attributes. One challenge in making thematic maps is to figure out which features to include as the minimally required locational reference information (that is, what to exclude from the map). In contrast, for general reference maps, the challenge is to figure out which classes of features are of greatest interest to a wide range of users (that is, what to include on the map).

Thematic maps can be used to communicate a wide range of data, including counts, amounts, ratios, and ranks. Knowing the number of people or things at a specific location can add a whole new dimension to comparing features. For example, knowing how many people live in each administrative region of a country might enable humanitarian planners to plan food distribution or long-term agricultural development programs.

Quantities can be mapped in one of three ways: as discrete features, as continuous phenomena, or as data summarized by area.

1. **Discrete features** are those that represent a single item or value. They can be points, lines, or areas, and the quantity associated with them is for that single feature. If the points represent food warehouses, for example, the quantity could indicate their maximum storage capacity. Or you may have lines that represent roads with a count of the number of days that they are serviceable each year. Or you may have areas that represent agricultural fields with an estimate of the volume of their expected harvest in bushels or tons.
2. **Continuous phenomena** are usually raster datasets that show a continuum, that is, a continuous range of values over a large area. Weather maps showing regional rainfall or temperature and hillshade maps revealing land slope and elevation are examples of maps displaying continuous phenomena.
3. **Data summarized by area** are polygon datasets communicating information that pertains to a defined area. In this exercise, we explore census data that summarizes the number of people living in administrative regions of Timor-Leste. You can probably think of many other examples of how demographic, economic, statistical, and other types of data are most effectively summarized by area.

UN special mission to assess need

Food security—having enough to eat—is a chronic challenge for many developing countries. Unfavorable agricultural conditions, dysfunctional markets, and civil unrest are just some of the problems that lead to unreliable, insufficient food supplies for vulnerable populations. In the southeast Asian country of Timor-Leste (East Timor), food insecurity is caused by a combination of factors, including poverty, poor infrastructure, high vulnerability to natural hazards, and very low levels of human development and productivity. The mass displacement of people and widespread destruction of critical infrastructure during its fight for independence between the late 1990s and 2002, combined with three consecutive years of drought (2001–2004), left this newly independent nation with one of the highest malnutrition rates in Asia (World Food Programme 2006).

In its June 2007 food security assessment of Timor-Leste, a special mission from the Food and Agriculture Organization (FAO) of the United Nations and the World Food Programme (WFP) found that extreme weather had once again severely affected production in the country. To make a bad situation even worse, locusts had consumed harvestable crops, which discouraged some farmers from replanting for the next season for fear of wasting their efforts. Maize, rice, and cassava were the worst-affected crops, with a nearly 30 percent decline in production compared to previous years (Food and Agriculture Organization of the United Nations and World Food Programme 2007, United Nations Office for the Coordination of Humanitarian Affairs 2008).

The FAO-WFP mission estimated that approximately 20 percent of the population would require emergency food assistance over the following six months (Food and Agriculture Organization of the United Nations and World Food Programme 2007). As of February 2008, an additional 23 percent were highly vulnerable to becoming food insecure (United Nations Office for the Coordination of Humanitarian Affairs 2008).

Scenario: Mapping food insecurity in Timor-Leste

In this chapter, you will produce a map that represents several themes of data related to food security in Timor-Leste. You will discover that there are countless ways to visualize spatial data—and to reveal key themes within that data. For the humanitarian community, these types of maps are essential to communicating both internally and with partner organizations, such as donor agencies, host governments, and nongovernmental aid organizations.

In the context of humanitarian emergencies, cartographic representation can be a powerful means of converting complex datasets into actionable information, allowing busy decision makers to quickly grasp overall trends and take appropriate action.

Study table 2.1 (pages 46–48). Even if you are not familiar with the June 2007 FAO–WFP report from which it was extracted, are you able to glean any key messages from the data? Caution must always be employed in interpreting data, but the table does provide a very interesting picture of food security in each district of Timor-Leste at the time. Note that the table describes aggregate, not per-capita, food availability.

District	Maize	Rice	Cassava and other root crops[b]	Total
Aileu				
Total availability	**2,448**	**419**	**1,085**	**3,952**
Production	2,448	419	1,085	3,952
Total utilization	**4,898**	**3,555**	**736**	**9,189**
Food use	4,221	3,517	703	8,441
Seed feed and losses	677	38	33	747
Deficit (surplus)	**2,450**	**3,136**	**(349)**	**5,237**
Ainaro				
Total availability	**4,001**	**743**	**1,750**	**6,495**
Production	4,001	743	1,750	6,495
Total utilization	**6,545**	**4,594**	**958**	**12,096**
Food use	5,432	4,527	905	10,864
Seed feed and losses	1,113	67	53	1,232
Deficit (surplus)	**2,543**	**3,850**	**(793)**	**5,601**
Baucau				
Total availability	**6,616**	**3,018**	**2,042**	**11,676**
Production	6,616	3,018	2,042	11,676
Total utilization	**13,910**	**10,388**	**2,083**	**26,380**
Food use	12,131	10,109	2,022	24,262
Seed feed and losses	1,779	279	61	2,119
Deficit (surplus)	**7,294**	**7,370**	**41**	**14,705**
Bobonaro				
Total availability	**3,718**	**4,074**	**2,512**	**10,304**
Production	3,718	4,074	2,512	10,304
Total utilization	**9,431**	**7,343**	**1,482**	**18,256**
Food use	8,440	7,033	1,407	16,879
Seed feed and losses	991	310	75	1,377
Deficit (surplus)	**5,712**	**3,269**	**(1,030)**	**7,951**
Cova Lima				
Total availability	**8,436**	**4,407**	**2,854**	**15,698**
Production	8,436	4,407	2,854	15,698
Total utilization	**7,641**	**4,843**	**986**	**13,469**
Food use	5,401	4,501	900	10,803
Seed feed and losses	2,239	341	86	2,666
Deficit (surplus)	**(795)**	**435**	**(1,869)**	**(2,229)**
Dili				
Total availability	**778**	**40**	**563**	**1,381**
Production	778	40	563	1,381
Total utilization	**10,657**	**8,701**	**1,756**	**21,114**
Food use	10,437	8,697	1,739	20,873
Seed feed and losses	220	3	17	241
Deficit (surplus)	**9,879**	**8,661**	**1,193**	**19,733**

District	Maize	Rice	Cassava and other root crops[b]	Total
Ermera				
Total availability	**2,460**	**618**	1,392	**4,469**
Production	2,460	618	1,392	4,469
Total utilization	**11,136**	**8,769**	1,784	**21,689**
Food use	10,456	8,713	1,743	20,912
Seed feed and losses	680	56	42	777
Deficit (surplus)	**8,676**	**8,151**	392	**17,219**
Lautem				
Total availability	**14,491**	**2,598**	4,883	**21,973**
Production	14,491	2,598	4,883	21,973
Total utilization	**9,903**	**5,241**	1,148	**16,291**
Food use	6,008	5,006	1,001	12,015
Seed feed and losses	3,895	234	146	4,276
Deficit (surplus)	**(4,588)**	**2,643**	**(3,735)**	**(5,681)**
Liquica				
Total availability	**1,400**	**236**	445	**2,081**
Production	1,400	236	445	2,081
Total utilization	**6,454**	**5,078**	1,025	**12,557**
Food use	6,068	5,057	1,011	12,136
Seed feed and losses	386	21	13	421
Deficit (surplus)	**5,054**	**4,842**	580	**10,476**
Manatuto				
Total availability	**3,508**	**3,027**	1,914	**8,448**
Production	3,508	3,027	1,914	8,448
Total utilization	**4,771**	**3,457**	694	**8,923**
Food use	3,819	3,183	637	7,639
Seed feed and losses	952	274	57	1,284
Deficit (surplus)	**1,263**	**431**	**(1,220)**	**474**
Manufahi				
Total availability	**6,612**	**2,519**	1,411	**10,542**
Production	6,612	2,519	1,411	10,542
Total utilization	**6,337**	**3,992**	801	**11,130**
Food use	4,552	3,793	759	9,103
Seed feed and losses	1,785	199	42	2,027
Deficit (surplus)	**(275)**	**1,473**	**(610)**	**588**
Oecusse				
Total availability	**2,976**	**1,440**	2,621	**7,037**
Production	2,976	1,440	2,621	7,037
Total utilization	**6,749**	**4,985**	1,050	**12,783**
Food use	5,826	4,855	971	11,652
Seed feed and losses	923	130	79	1,131
Deficit (surplus)	**3,773**	**3,545**	**(1,571)**	**5,746**

District	Maize	Rice	Cassava and other root crops[b]	Total
Viqueque				
Total availability	**12,376**	**3,600**	**3,437**	**19,413**
Production	12,376	3,600	3,437	19,413
Total utilization	**10,445**	**6,222**	**1,288**	**17,956**
Food use	7,112	5,927	1,185	14,224
Seed feed and losses	3,334	295	103	3,732
Deficit (surplus)	**(1,931)**	**2,622**	**(2,148)**	**(1,457)**
[a]Source: FAO-WFP 2007. [b]Cereals equivalent.				

Table 2.1   Timor-Leste food balance in tons by district in 2007[a]

Even a quick look reveals that more districts were experiencing deficit production than surplus production and that the district of Dili was the most food insufficient. The values for production are identical to the values for total availability for every district, which suggests that other forms of food availability (external aid, past production inventories, etc.) either were not counted or already depleted. Such a table could be helpful to decision makers at the headquarters level as well as at the field level. Yet in order to adequately interpret the data, decision makers may prefer it represented geographically.

Therefore, you are going to produce a multitheme poster containing four different types of thematic maps. Through the exercises that follow, you will use the preceding tabular data, together with other data layers, to cartographically communicate the conditions of food insecurity in Timor-Leste in 2007.

Table 2.2 lists the relevant attributes of each data layer used in this chapter.

Layer[a] or attribute	Description
Districts.shp	Timor-Leste second administrative layer boundaries (polygons)
DIST_NAME	District name
POP_2001	District population (2001 census)
SubDistricts.shp	Timor-Leste third administrative layer boundaries (polygons)
SUB_NAME	Subdistrict name
POP_2001	Subdistrict population (2001 census)
Sucos.shp	Timor-Leste fourth administrative layer boundaries (polygons)
SUCO_NAME	Suco name
POP_2001	Suco population (2001 census)
Towns.shp	Timor-Leste towns (points)
NAME	Town name
TYPE	Capital, major town, or minor town
CropForecast2007.dbf	Timor-Leste crop forecast by district for 2007
DIST_NAME	District name
MAIZE	Maize forecast
PADDY	Paddy rice forecast
CASSAVA	Cassava forecast
OTHER_ROOT	Other crop forecast
TOTAL	Total crop forecast

Layer[a] or attribute	Description
FoodSupplyData.dbf	Timor-Leste food supply by district for 2007
DIST_NAME	District name
TOTAL_PROD	Total food production
TOTAL_USE	Total food consumption
BALANCE	Total food surplus or deficit
Roads.shp	Timor-Leste paved roads (lines)
Indonesia.shp	Indonesia country polygon
[a]Files with the .dbf extension are database tables.	

Table 2.2 Data dictionary

Exercise 2.1

Displaying data

In this exercise, you will explore census data that summarizes the number of people living in an area, which is an example of the third type of quantitative data. You can probably think of many other examples of how demographic and other types of data are most effectively summarized by area.

Use graduated symbols to illustrate quantitative attribute data

1. In ArcMap, open GISHUM_C2E1.mxd.

Your map display shows the southeast Asian country of Timor-Leste, with some of its road network and towns. A gray mask represents its neighbor, Indonesia. You can also see polygons delineating the second administrative level (districts), the third administrative level (subdistricts), and the fourth administrative level (*sucos*). The districts are labeled and all areas are uniformly colored.

You will now recolor the map to display the population value of each suco.

2. In the table of contents, right-click the Sucos layer.

3. Select Open Attribute Table.

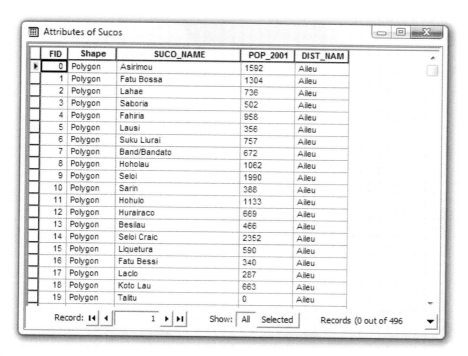

The Attributes of Sucos table contains the names of the 496 sucos in Timor-Leste, their estimated populations in 2001, and the names of their districts. You will use the field POP_2001 to symbolize the Sucos layer.

4.   **Close the Sucos attribute table.**

5.   **Right-click the Sucos layer, and then open Properties.**

6.   **In the Layer Properties dialog box, click the Symbology tab (if it's not the default).**

7.   **In the Show box on the left, select Quantities > Graduated colors.**

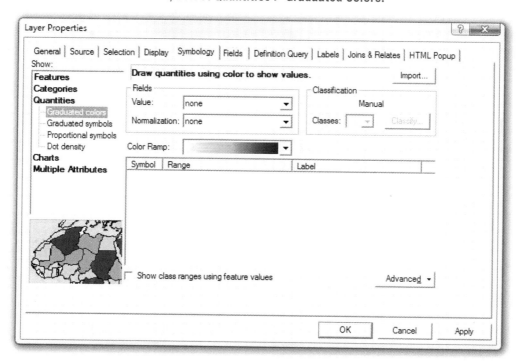

8. **In the Fields area, set the Value field to POP_2001.**

9. **Click the drop-down arrow next to the color ramp, scroll through the various options, and choose the color ramp that ranges from light orange to dark orange.**

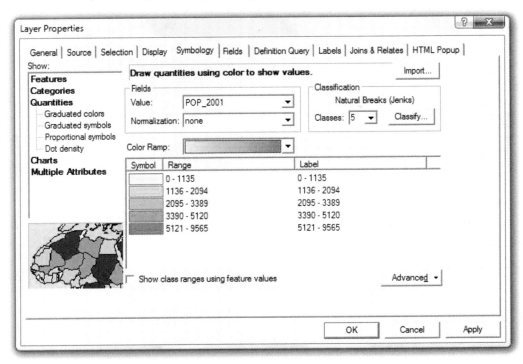

The Sucos layer has been assigned five number classes by default, with darker shades corresponding to incrementally higher population counts.

The most common method of grouping values into categories (or classes)—and the default choice in ArcMap—is the natural breaks (Jenks) method. Classes are based on natural groupings inherent in the data. ArcMap identifies break points that maximize the differences between classes. Feature class boundaries are set where there are relatively big jumps in data values.

ArcMap provides seven classification schemes for grouping data: manual, equal interval, defined interval, quantile, natural breaks (Jenks), geometrical interval, and standard deviation. Choosing the best classification method requires a good understanding of the distribution of your data. Evenly distributed data might be a candidate for a quantile or equal-interval classification. Data that is tightly grouped, or for which you want to show the median values, might be a better candidate for a standard deviation classification. Data that has one or more distinct groupings might be best shown using Jenks natural breaks or a manual classification.

After choosing a method of classification, decide how many classes of values will allow you to communicate most effectively with your target audience—too many classes can make a map difficult to read; too few can oversimplify the data. Highly sophisticated audiences may appreciate a large number of classifications. Busy, nontechnical decision makers usually prefer few classes that enable them to quickly glance at a map and understand the general split between low, medium, and high values of any theme. Therefore, three to five quantity classes often work best for general audiences.

10. **Click the Classify button in the upper right of the Layer Properties window to open the Classification dialog box.**

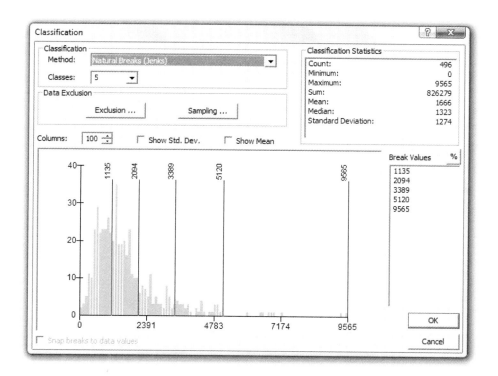

The data values are bunched up on the left side of the chart, and the gaps between the classification break lines in blue get larger as the values increase. Viewing this on the map, you would expect to see a large number of polygons representing the first two classes and just a few representing the rest of the data.

11. **Click the Classification Method drop-down list, select each method in turn, and observe how different methods affect the classification histogram.**

12. **Change the classification method to Manual.**

13. **Click the first value in the Break Values box on the right.**

At the bottom of the dialog box, you can see that there are 197 features in the first class.

14. **Change the first break value to 1000.**

Notice there are now 160 features in the first class.

15. **Change the second value to 2000, the third value to 3000, and the fourth value to 4000. Leave the fifth value at 9565, since that represents the upper bounds of the largest population class.**

By choosing rounded population values, you are making it simpler for the target audience to appreciate the relative population of each suco.

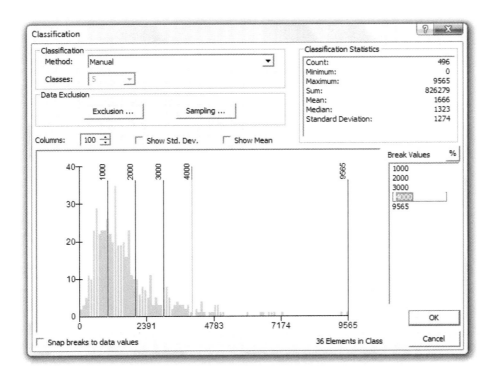

16. Click OK once your Classification dialog box matches the preceding image.

17. Click OK to close the Layer Properties dialog box and apply the new symbology settings.

Now your map of Timor-Leste shows each suco's population range: 0–1000, 1001–2000, 2001–3000, 3001–4000, and 4000 and above. This use of shaded polygon symbology is called **choropleth mapping.**

Use marker symbols to illustrate qualitative attribute data

As you learned in the previous chapter, ArcMap allows you to customize the design and properties of point markers (as well as lines, polygons, and annotation). Since your map of Timor-Leste includes locations of towns, you are going to select symbols that identify the national capital, major towns, and minor towns.

1. **Right-click the Towns layer, and then select Properties.**

2. **In the Show box on the Symbology tab, select Categories > Unique values.**

3. **Set the Value Field to TYPE.**

4. **Click the Add All Values button.**

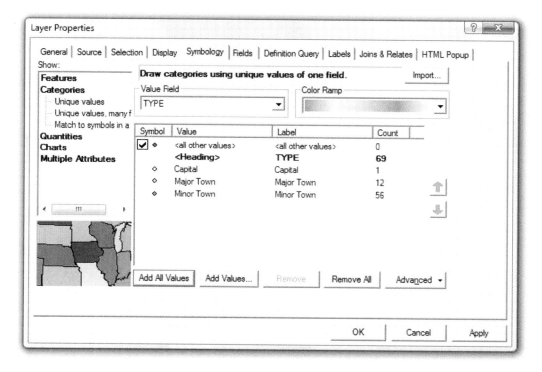

The features in the Towns layer are categorized by type: there is one capital, 12 major towns, and 56 minor towns (you can see this if you open the attribute table for Towns).

5. **Double-click the default symbol for Capital to open the Symbol Selector dialog box.**

6. **Select the Star 1 symbol, and then change its size to 22.00.**

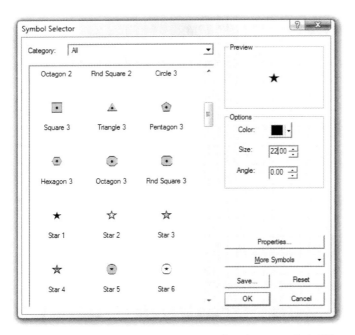

7. **When your settings match the preceding image, click OK.**

8. **Double-click the symbol for Major Town, and then select the Circle 2 symbol.**

9. **Change the circle's size to 9 and its color to Mars Red, then click on the Properties button.**

You are now going to edit the symbol properties a little further to create a white halo around the red circle.

10. **In the Symbol Property Editor dialog box, change the outline layer color of the symbol for major towns to Arctic White.**

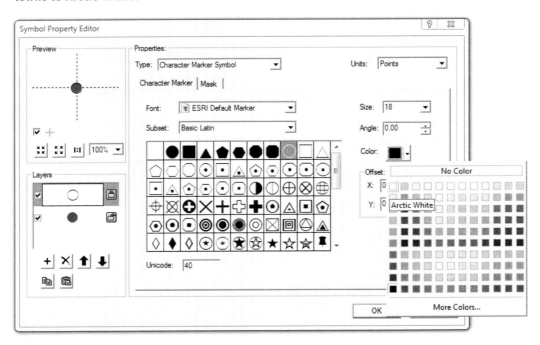

11. Click OK, and then click OK again to return to the Layer Properties dialog box.

12. Using the same process as above, change the symbol for Minor Town to Circle 2, with a size of 7, a fill color of Medium Apple green, and an outline color of Arctic White.

Symbol	Value	Label	Count
☑ ●	<all other values>	<all other values>	0
	<Heading>	**TYPE**	**69**
∘	Minor Town	Minor Town	56
●	Major Town	Major Town	12
★	Capital	Capital	1

13. When your symbols match the preceding image, click OK on the Layer Properties dialog box to return to the map display.

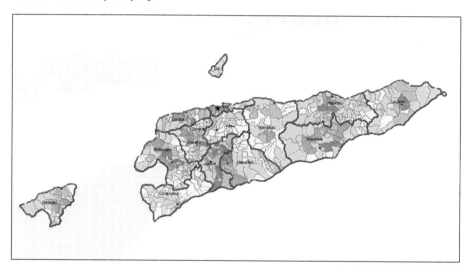

The point marker symbols that you have chosen for the capital and the major and minor towns of Timor-Leste now appear on the map and in the table of contents.

14. Click Save As from the File menu, and then name your map document **GISHUM_C2E2.mxd** in preparation for the next exercise.

About saving map documents

After you finish working on a map, you can save it, exit ArcMap, and return to your exact settings the next time you open ArcMap.

If you haven't saved the map before, you'll need to name it, preferably with a name that adequately describes its contents. ArcMap automatically appends a file extension (.mxd) to your map document name. You can also save your map as a map template (.mxt).

The data displayed on a map is not saved with it. Map layers reference the data sources in your GIS database. This helps keep map documents relatively small in size. So if you plan to distribute your map to others, they'll need access to both the map document and the data your map references.

It is always a good idea to save your map periodically while editing it. You can select Save As from the File menu if you wish to save it as a new file, which will preserve your original map document.

What to turn in

If you are working in a classroom setting with an instructor, submit an electronic copy or a printout of your map.

Exercise 2.2

Joining tabular data with spatial data

Up to now, you have used values within the attribute tables of spatial data layers to symbolize and classify your maps. Frequently you will need to map data that is stored in databases or spreadsheets. In order to do this, you must join the nonspatial data with a spatial data layer using some common field, such as a place-name or postal code.

First, let's load the tabular data that you reviewed at the beginning of the chapter into our table of contents.

Perform a join

1.   **If you are starting a new session, open GISHUM_C2E2.mxd.**

2.   **On the Standard toolbar, click the Add Data** ➕ **button.**

3.   **Navigate to your Chapter2 folder, and then add the FoodSupplyData.dbf file.**

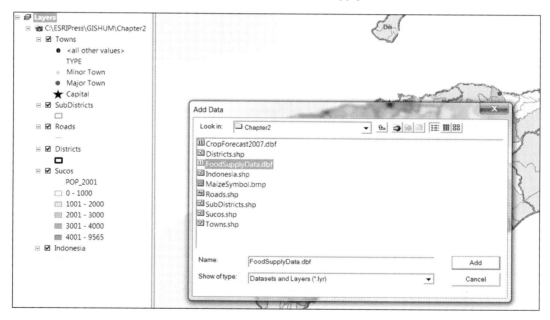

4.   **Right-click FoodSupplyData in the table of contents, and then select Open.**

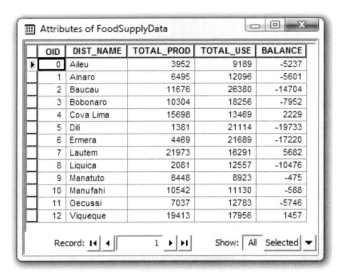

The table lists the amounts of total production (TOTAL_PROD) and consumption (TOTAL_USE) of food (cereals) for each district in Timor-Leste. The BALANCE column lists the differences between production and consumption, with negative values indicating shortfalls.

5. **Move the FoodSupplyData table to the right side of your screen.**

6. **Open the attribute table of the Districts layer and compare it with the FoodSupplyData table.**

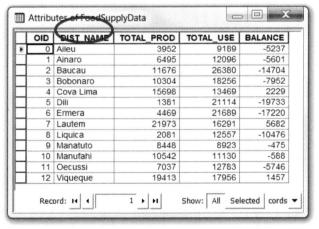

The two tables share a common field, DIST_NAME, and therefore you can use the Districts layer to spatially project the FoodSupplyData table onto your map display. Because the joining of tables requires that the records in the common field be identical, alphanumeric codes are usually preferred, since place-names are often inconsistently spelled (especially for regions where Latin characters are not commonly used or where multiple languages and dialects are spoken). However, since we do not have a uniform place code in our dataset, we will employ the district names, as it appears that they are consistently spelled in the two tables.

7. **Close the FoodSupplyData and Districts attribute tables.**

8. **Right-click the Districts layer, and then select Joins and Relates > Join.**

9. **Populate the Join Data dialog box as follows:**

- What do you want to join to this layer: Join attributes from a table
- Choose the field in this layer that the join will be based on: DIST_NAME
- Choose the table to join to this layer: FoodSupplyData
- Choose the field in the table to base the join on: DIST_NAME
- Join Options: Keep all records

ArcMap is fairly clever at guessing the settings that may be needed in many types of operations, based on the data that has been loaded into the current session.

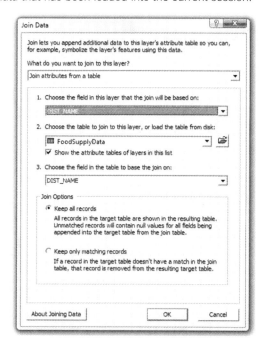

10. When your Join Data settings match the preceding image, click OK.

Let's check to ensure that the table join was successful.

11. Open the Districts attribute table. You may need to expand your table to see the additional fields created through the join to the FoodSupplyData table.

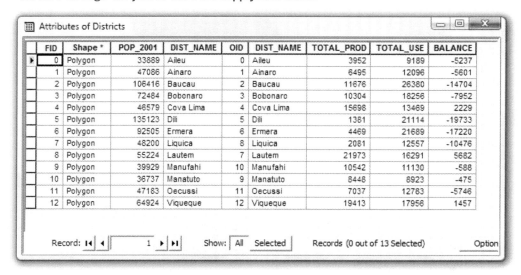

You are now ready to project the food supply data onto your map display.

12. Close the Districts attribute table.

Illustrate joined tabular data using graduated symbols

1. Right-click the Districts layer, and then select Properties.

2. On the Symbology tab, choose Quantities > Graduated colors.

3. Change the Value Field to TOTAL_PROD.

Notice the additional field values that appear in the drop-down list due to the join between the Districts layer and FoodSupplyData table.

4. Change the color ramp to the one that ranges from light to dark purple.

5. Click the Label column header, and then select Format Labels.

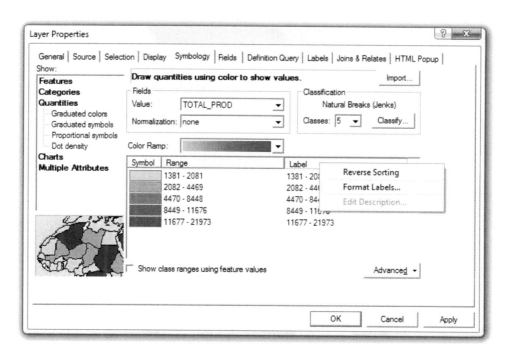

6. **Set the Number Format dialog box parameters as follows:**

    - Category: Numeric
    - Rounding: Number of decimal places (0)
    - Alignment: Left
    - Show thousands separators
    - Pad with zeros

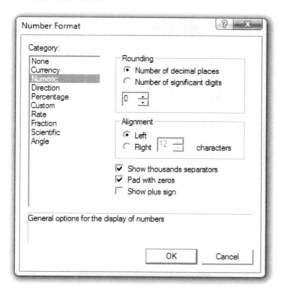

7. **When your Number Format settings match the preceding image, click OK.**

The values in the food production category labels will now have comma separators.

8. **Click OK to close the Layer Properties dialog box.**

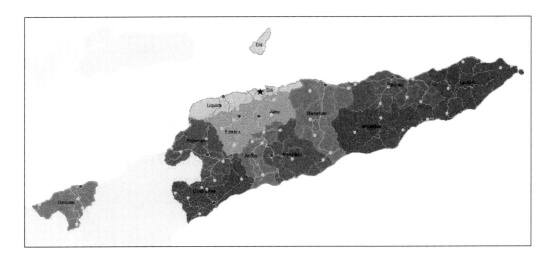

You now have a choropleth map of food production by district. Because the Districts layer's symbology has changed, the Sucos layer is hidden and the district boundaries are less visible.

You will now copy the Districts layer and restore its original boundary symbology.

9.   Right-click the Districts layer, and then select Copy.

10. Right-click the Layers dataset, and then select Paste Layer(s).

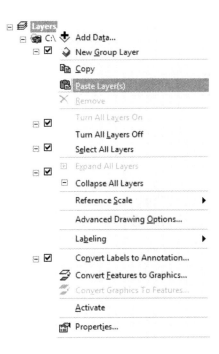

The Districts layer will be duplicated in the table of contents. You will now modify its settings to simply provide a black outline to each district in the map display.

11. Right-click the copied layer (it is the first of the two Districts layers in the table of contents), and then select Joins and Relates > Remove Join(s) > Remove All Joins.

12. Right-click the copied Districts layer, and then select Properties.

13. Go to the General tab, and then rename the layer **District Boundaries**.

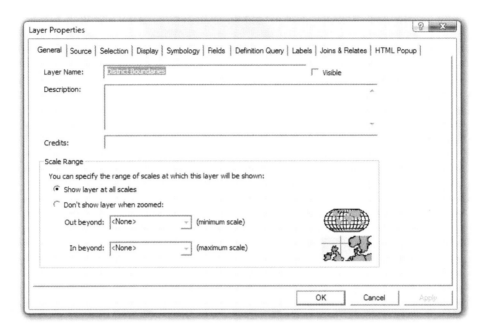

14. Go to the Symbology tab. In the Show box, select Features > Single symbol (it may already be selected for you).

15. Click the Symbol button, and then change the default settings to no fill color, outline width 2, and outline color black.

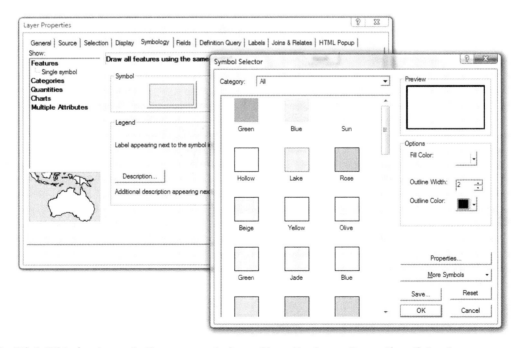

16. Click OK twice to apply the new symbology. Close the Layer Properties dialog box.

The District Boundaries layer now appears in the map display, but it is blocking the SubDistricts and Roads symbols. The hierarchy of layers can be changed to optimize the map display.

17. **Click the Display tab at the bottom of the table of contents, and then demote the District Boundaries layer by clicking and dragging it below the Roads layer and above the Districts layer.**

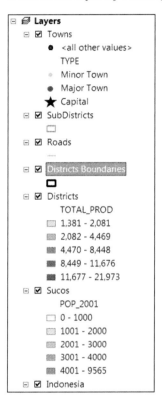

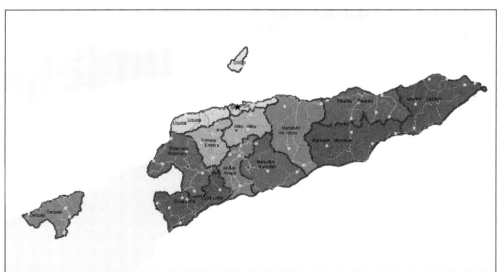

Note that because Label Features is turned on for both the District Boundaries layer and the Districts layer, the names of each district are duplicated in your map display.

18. **Click Save As from the File menu, and then name your map document GISHUM_C2E3.mxd in preparation for the next exercise.**

What to turn in

If you are working in a classroom setting with an instructor, submit an electronic copy or a printout of your food production choropleth map.

Your turn

Experiment with the different color ramps provided in ArcMap. Consider the significance of the color, range, and order when illustrating various types of phenomena. What color ramp would you select to reflect a region's elevation? What color ramp would you select to reflect a city's exposure to toxic hazards?

Exercise 2.3

Selecting data by attribute value

A very common type of database query is to select features according to their attributes. In this exercise you will use ArcMap to identify every district that has a food deficit greater than or equal to 5,000 tons. You will then export those selected records to a separate table that identifies districts with such significant levels of food insecurity.

1. **If you are starting a new session, open GISHUM_C2E3.mxd.**

2. **On the main menu, click Selection > Select By Attributes.**

3. **In the Select By Attributes dialog box, choose the following settings:**

    - Layer: Districts
    - Method: Create a new selection

You are now ready to select districts with a food deficit of 5,000 or more tons.

4. **Scroll down the list of Districts fields, and then double-click "FoodSupplyData.BALANCE" to add it to the query expression box.**

If you want to sort the list of fields or see the fields by their aliases, click the small button on the top right of the fields list. By default, ArcMap prefixes the field names with the name of the joined layer in which they are contained.

5. **Click the <= operator to continue building your query expression.**

6. **Now type -5000, so that the expression reads "FoodSupplyData.BALANCE" <= -5000.**

Whenever you are building an SQL (Structured Query Language) expression in ArcGIS, you have the option to click Verify to ensure that your expression is viably constructed and has correct syntax. If there is a problem, click Clear to build the expression again or click Help if you need assistance.

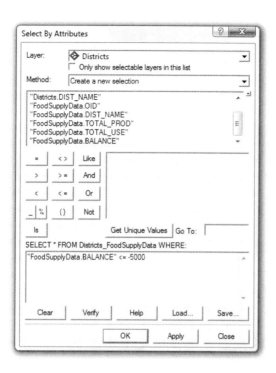

7. **Once your settings match the preceding image, click OK.**

ArcMap will highlight districts that have been selected according to the criteria.

8. **To see the name and attribute details of selected districts, open the attribute table of the Districts layer.**

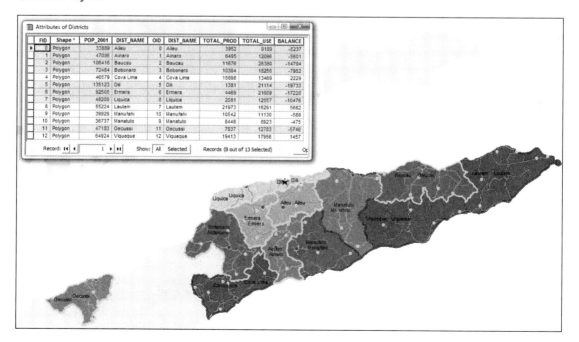

The eight districts with food deficiencies of 5,000 tons or more are highlighted in the table.

9. On the bottom bar of the attribute table, click the Selected tab to limit the display to only those records that are currently selected.

10. To export this information from ArcMap (to be used in Microsoft Excel, Access, etc.), click the Options button at the bottom right of the Districts attribute table and select Export.

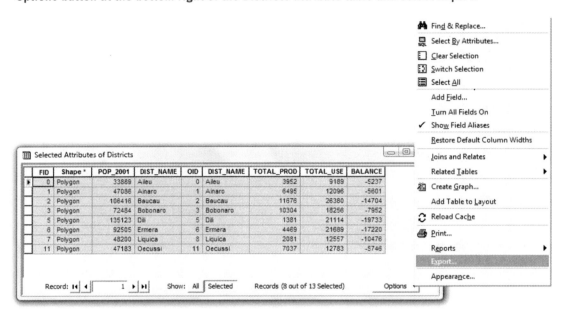

11. In the Export Data dialog box, choose Selected records, name the output table **FoodDeficit5000Plus** to be saved in your Chapter2 folder, and click OK.

12. Click No when asked if you want to add the table to your current map.

13. Click the Clear Selected Features  button on the Tools toolbar to cancel the selection from your ArcMap session (this can also be done by going to the main menu and clicking Selection > Clear Selected Features).

14. Close the attribute table.

15. **If you wish to save your work, click Save As from the File menu, but choose a name other than GISHUM_C2E4.mxd (which already exists and should not be overwritten).**

What to turn in

If you are working in a classroom setting with an instructor, submit an electronic copy of your FoodDeficit5000Plus.dbf and a screen capture or printout of your GISHUM_C2E3.mxd map.

Exercise 2.4

Mapping data

Navigate data frames using Layout View

1. **Open GISHUM_C2E4.mxd.**

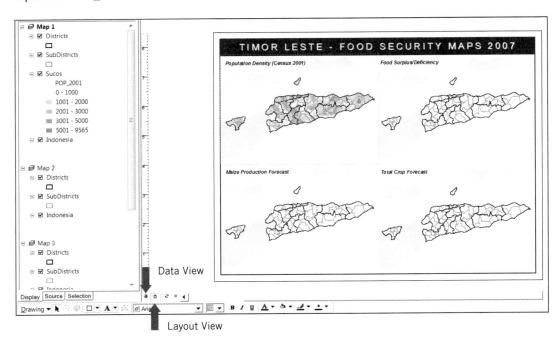

You are currently in layout view, which we briefly explored in chapter 1. Layout view provides a virtual page upon which geographic data and map elements are placed and arranged for printing. In comparison, data view is for exploring, displaying, and querying geographic data, and therefore it does not incorporate titles, legends, north arrows, scale bars, and other map elements.

Layout view contains a template with four inset maps, or data frames, of Timor-Leste. In this exercise, you will modify these data frames using different thematic mapping techniques.

The map in the top left (Map 1 in the table of contents) is similar to the one you created in exercise 2.1; it is activated because your first task is to show population with dot density symbols instead of the graduated quantitative symbols (shaded polygons) currently showing. The other three maps still need to be generated, and those data frames simply provide placeholders for you to modify.

Before working on the first map, let's take a moment to learn how to navigate between data view and layout view, and practice creating and removing data frames.

2. **Click the Data View tab to switch from the layout view to the data view of Map 1.**

3. **Now click the Layout View tab, and then click the Food Surplus/Deficiency map.**

Notice that Map 2 is now activated. If you return to the data view, any modifications you make will be applied only to Map 2 (you can also activate a data frame in data view by right-clicking the desired map in the table of contents and selecting Activate).

New data frames can be added to your map document by going to the main menu and clicking Insert > Data Frame.

Create a dot density map (Map 1: Population Density)

Although the current choropleth is quite effective, you are going to resymbolize the population of each suco in Timor-Leste using a method called "dot density mapping," with dot density representing relative population density.

1. **Switch to data view.**

2. **Right-click Map 1 in the table of contents, and then select Activate.**

3. **Open Properties for the Sucos layer.**

4. **In the Show area on the Symbology tab, click Quantities > Dot density.**

5. **In the Field Selection box, click POP_2001 and then the Add (>) button to add POP_2001 as the field for generating the dot densities.**

The Symbology editor will analyze the field values and produce default densities and presentation settings for you to apply or modify.

6. **In the Symbol column in the right panel of the Field Selection window, right-click the default dot symbol (it may appear faint), and then change its color to Mars Red.**

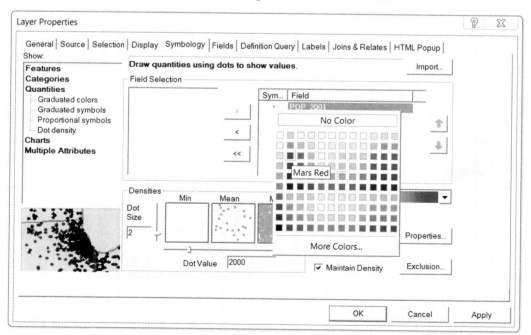

7. Click the **Background line symbol** on the lower right of the Layer Properties box, and then remove the suco outlines by clicking **No Color** in the the Symbol Selector dialog box.

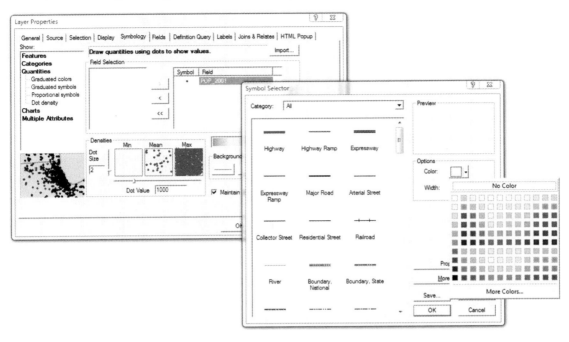

8. Click **OK** to close the Symbol Selector dialog box.

9. Now click the **Background fill** drop-down arrow, and then change the suco fill color to **Yucca Yellow**.

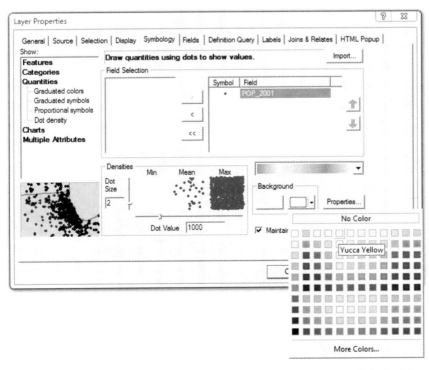

10. In the Densities area, maintain the default dot size, and then click the **Mean density option**.

11. Change Dot Value to 500.

The Population Density map will now use one dot to represent 500 people.

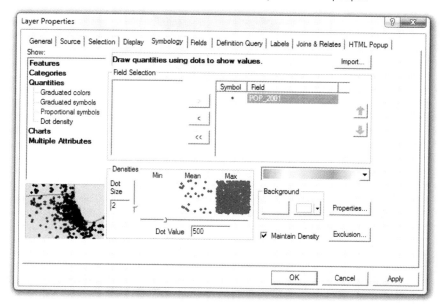

12. When your settings match the preceding image, click OK.

Your Population Density map should now look similar to the following map.

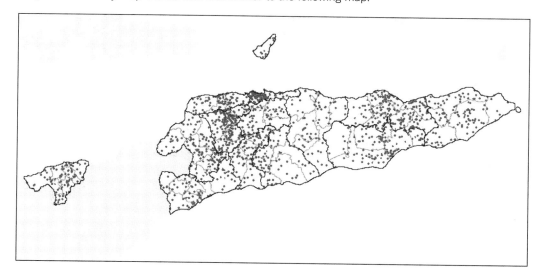

The placement of dots is random within each feature of a layer. In our map, the arrangement of dots does not match the locations of populations within each suco. While dot density maps can reveal meaningful overall patterns, the random distribution of dots can be misleading.

13. Click Save on the Standard toolbar to update your map document.

Create a choropleth map (Map 2: Food Surplus/Deficiency)

You will now shift your attention to the next map and generate choropleths (shaded polygons) to represent areas that have food surplus and food deficiency.

1. **Activate Map 2, and then go to data view.**

2. **Add the FoodSupplyData.dbf from your exercise folder to Map 2's table of contents.**

3. **Using the process employed in exercise 2.2, join the FoodSupplyData table to the Districts layer.**

4. **Open the Districts layer's properties, and then change the default symbology to Quantities > Graduated colors.**

5. **Set the Fields Value to BALANCE.**

6. **Click Classify to open the Classification dialog box.**

7. **Change the Classification method to Defined Interval, and then change Interval Size to 5000.**

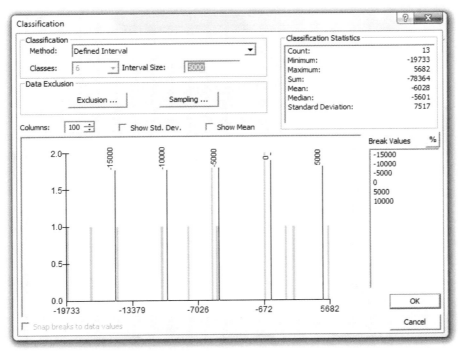

The defined-interval method automatically determines the number of classes that are required to divide the range of food balance values into classes (groups) of 5,000.

8. **Click OK to close the Classification dialog box, and then apply the red-to-green color ramp to emphasize those districts that are most food deficient. You can customize the color ramp by double-clicking the symbol for each range to change the default ramp colors.**

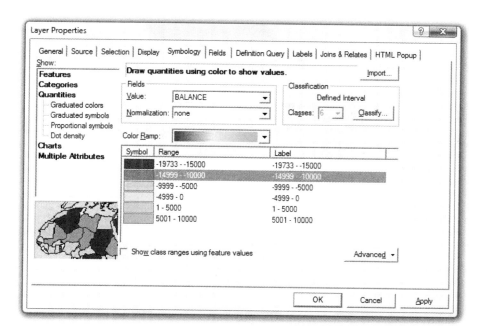

9. When your settings match the preceding image, click **OK**.

Although your map now shows the ranges of food deficiency and surplus for each district in Timor-Leste, it is not very useful to the reader without district names or a legend that explains the symbol ranges. We will add legends to all of the data frames in the next exercise, but we can add the district labels before editing the next map.

10. **Right-click the Districts layer, and then select Label Features.**

11. **Since the labels are fairly small, open the Districts layer's properties and, on the Labels tab, change the Text Symbol settings to Arial 10 Bold.**

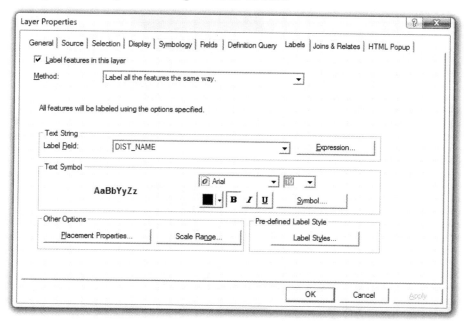

12. **Click OK to apply your new label settings.**

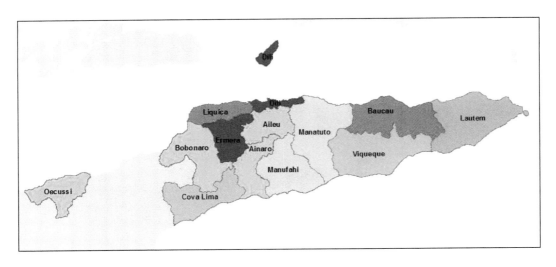

13. **Click Save on the Standard toolbar to update your map document.**

Create a graduated symbol map (Map 3: Maize Production Forecast)

You are now ready to start working on Map 3, which depicts the 2007 maize production forecast for Timor-Leste. You will use graduated symbols to represent the anticipated maize harvest for each district. Graduated symbols are often used to represent point data, such as city population, but can also be used instead of choropleths to symbolize polygon data.

1. **Activate Map 3 and go to data view.**

2. **Add the CropForecast2007.dbf from your exercise folder to Map 3's table of contents.**

3. **Open the CropForecast2007 table.**

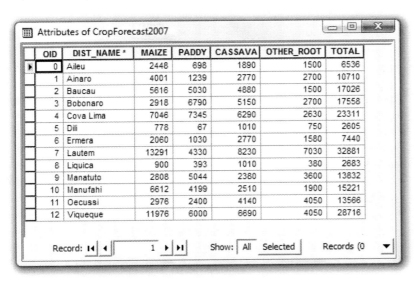

OID	DIST_NAME *	MAIZE	PADDY	CASSAVA	OTHER_ROOT	TOTAL
0	Aileu	2448	698	1890	1500	6536
1	Ainaro	4001	1239	2770	2700	10710
2	Baucau	5616	5030	4880	1500	17026
3	Bobonaro	2918	6790	5150	2700	17558
4	Cova Lima	7046	7345	6290	2630	23311
5	Dili	778	67	1010	750	2605
6	Ermera	2060	1030	2770	1580	7440
7	Lautem	13291	4330	8230	7030	32881
8	Liquica	900	393	1010	380	2683
9	Manatuto	2808	5044	2380	3600	13832
10	Manufahi	6612	4199	2510	1900	15221
11	Oecussi	2976	2400	4140	4050	13566
12	Viqueque	11976	6000	6690	4050	28716

The table lists the estimated 2007 harvest amounts for maize, paddy (rice), cassava (plantain), and other root crops, and the total crop harvest by district.

4. Close the table and, using the process employed earlier, join it to the Districts layer.

5. Open the Districts layer's properties and go to Symbology.

6. Change the default symbology to Quantities > Graduated symbols.

7. Set the Fields Value to MAIZE.

8. Change the classification method to defined intervals of **4000**.

While we could employ the default simple point marker for our graduated symbol, the map is intended for a general audience that might appreciate a more obvious symbol to illustrate the expected maize supply.

9. Click the Template button on the right of the Layer Properties dialog box, and then click Properties.

10. In the Symbol Property Editor, change the Type to Picture Marker Symbol.

11. Browse to the MaizeSymbol picture file in the Chapter2 folder and click Open.

The image of an ear of maize will now be used as the graduated symbol.

12. Click OK twice to return to Layer Properties.

13. Set the Districts layer's background fill color to Yucca Yellow, outline width to 1, and color to Gray 70 percent.

14. Set Symbol Size from 10 to 25 so that the four classification ranges can be easily distinguished.

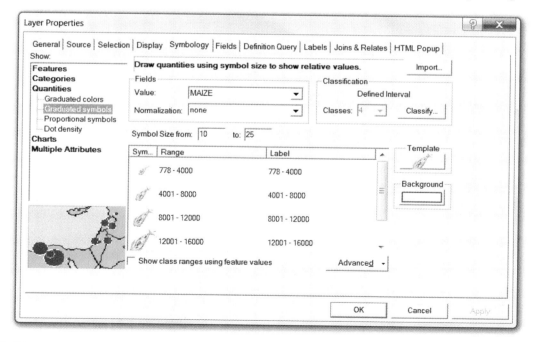

15. When your settings match the preceding image, click OK.

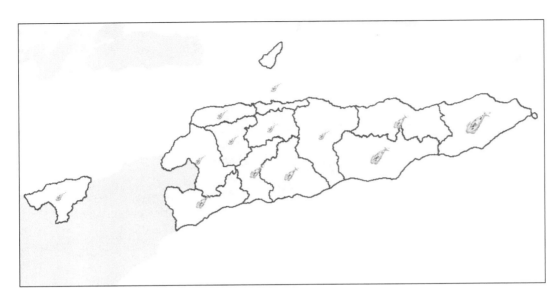

Graduated symbols are placed at the center of each polygon. Because some of Timor-Leste's districts are irregular in shape, their symbols fall along district boundaries or, in the case of the district of Dili, in the strait between the mainland and the island to the north. For the sake of expediency, we will accept this outcome, recognizing that perfect cartography is both an art and a science and takes much practice.

16. Click Save on the Standard toolbar to update your map document.

Create a bar chart map (Map 4: Total Crop Forecast)

The final type of thematic map you will create uses a bar chart to compare the various crop forecasts contained in the CropSupplyForecast2007 table. ArcMap provides three types of charts: pie charts, bar charts, and stacked charts. Pie charts represent each category of field values as a percentage—a complete pie chart represents 100 percent of all values. Bar charts, on the other hand, allow you to show amounts rather than just percentages. Each bar represents the value of each category in absolute terms so that comparisons can be made by the height of each bar. Stacked charts also show values, but are additive so that the column height is representative of the sum of all the field values for all categories.

1. **Activate Map 4, and then go to data view.**

2. **Add the CropForecast2007.dbf from your exercise folder to Map 4's table of contents.**

3. **Join the table to the Districts layer.**

4. **Open the Districts layer's properties, and then change the default symbology to Charts > Bar/Column.**

Slightly different than the Quantities options, the Charts options allow you to incorporate information from several different fields at once.

5. **In the Field Selection box, add the MAIZE, PADDY, CASSAVA, and OTHER_ROOT fields to the bar chart.**

6. **In the Color Scheme option, apply the natural-tones color ramp, which is the last ramp option.**

Since the color ramp assigns each field symbol a color randomly, you may wish to customize the color selections.

7. **Set the background color to Olivine Yellow (top row) for contrast against the color ramp.**

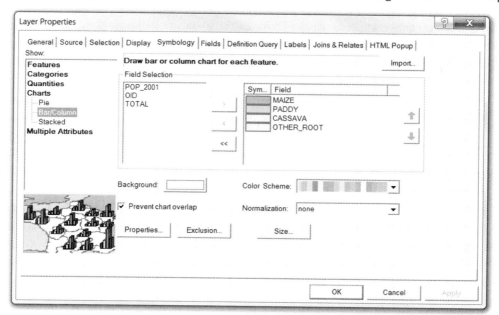

8. **When your settings match the preceding image, click OK.**

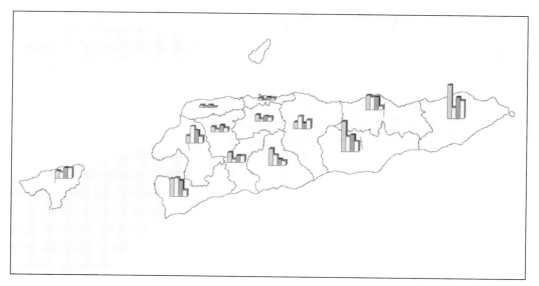

9. **Click Save As from the File menu, and then name your map document GISHUM_C2E5.mxd in preparation for the next exercise.**

Your turn

You may have noticed that the bar chart provides an option to normalize your values. Normalization is the process of dividing numeric attribute values by some common value to standardize (or "normalize") a dataset.

Normalize (divide) each district's forecast for maize, paddy, and so forth, by its total crop forecast to calculate the proportional distribution of each district's output by crop type. Note how the relative crop production is now emphasized in the map. Normalizing each district's forecast by the total crop forecast as a percent of the total for all districts does the same thing at a national level.

Exercise 2.5

Printing a multimap poster

You are now ready to put the final touches on your map document and then export it for large-format printing.

1. **If you are starting a new session, open GISHUM_C2E5.mxd.**

2. **Switch to layout view to see all four maps created in the previous exercise.**

3. **Activate the Population Density map.**

4. **On the main menu, go to Insert > Legend.**

The Legend Wizard will open, and all of the Map 1 layers will be added to the map's legend by default. Since our main purpose is to illustrate population density at the suco level, you will shorten the legend by eliminating items that are not relevant.

5. **Hold down the Ctrl key to highlight the Districts, SubDistricts, and Indonesia layers in the Legend Items panel, then click the Remove button.**

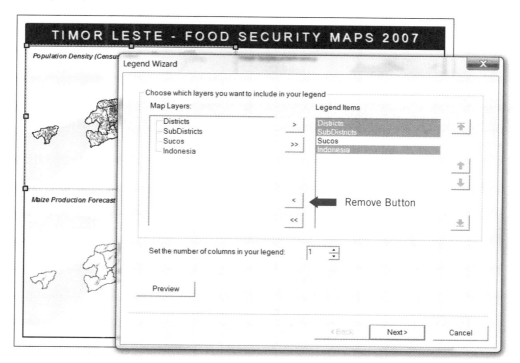

6. **When Sucos is the only layer listed under Legend Items, click Next.**

7. **Since the Legend Title does not add any value, delete "Legend" from the text box, and then click Next.**

8. **Click Next to skip the Legend Frame options, and then click Next again to skip the symbol patch options.**

9. Click Finish to accept the Legend settings and to add the legend to the layout.

10. Use the mouse to position the legend frame in a suitable location within the Population Density map frame.

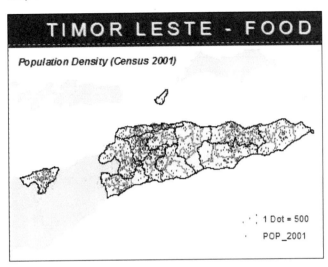

Your turn

Add a legend to each of the other three data frames, doing your best to optimize the legend properties for each thematic map.

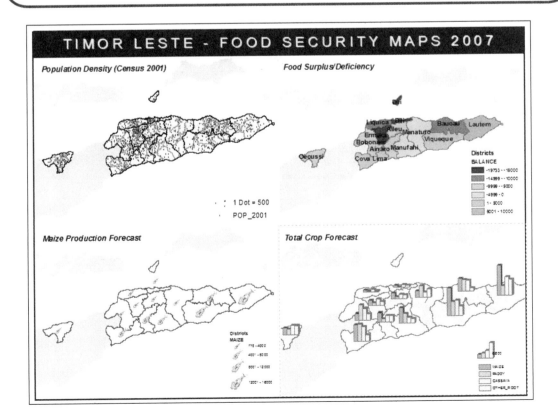

You are now ready to export your map to share with others or to send for output to a large-format printer.

11. **On the main menu, go to File > Export Map.**

12. **Name the map 2007FoodSecurityMap_TimorLeste to be saved in your Chapter2 folder.**

13. **Save the map as a PDF file type with a resolution of 300 dpi and Best (1:1) quality setting.**

You have completed the Timor-Leste food security mapping project. Before proceeding, take a moment to review what you accomplished in this chapter. After learning about the food security conditions of Timor-Leste, you used ArcGIS to import spatial data and symbolize that data employing various types of graphic representation. You then combined that spatial data with tabular data to produce a multithemed map display designed to communicate the 2007 food security conditions of Timor-Leste.

Do you now have a better sense of the potential of ArcGIS in helping people understand a humanitarian emergency and how to direct relief operations? In highly dynamic situations, decision makers rarely have time to study detailed reports or large tables of survey data. But a well-produced map can tell them what is happening in an instant.

What to turn in

If you are working in a classroom setting with an instructor, submit an electronic copy or printout of your 2007FoodSecurityMap_TimorLeste.pdf multimap poster display.

Assignment

Produce additional analyses of thematic, temporal, and spatial characteristics of food insecurity in Timor-Leste. Export the finished map sheet(s) complete with legends, titles, and so on, to a brief analytical report on food availability in Timor-Leste. Use sources such as the FAO–WFP report and journal articles to inform the reader of the historical and environmental context of the emergency. Also, review the actions taken by local and international organizations in response to the food production shortfalls, as well as the challenges they encountered.

What to turn in

If you are working in a classroom setting with an instructor, submit an electronic copy or printout of your report.

References

Fang, Cheng, Sharma Sharma, Raphy Favre, and Siemon Hollema. 2007. FAO/WFP food security mission to Nepal. Special report. Rome: Food and Agriculture Organization of the United Nations, World Food Programme.

Food and Agriculture Organization of the United Nations and World Food Programme. 2007. FAO/WFP crop and food supply assessment mission to Timor-Leste. Special report. Rome: United Nations.

United Nations Office for the Coordination of Humanitarian Affairs. 2008. http://www.irinnews.org. IRIN. February 4. http://www.irinnews.org/Report.aspx?ReportId=76552 (accessed March 3, 2008).

World Food Programme. 2006. Timor Leste: Comprehensive food security and vulnerability analysis (CFSVA). Rome: United Nations World Food Programme.

Chapter 3

Developing spatial data management skills

Disciplined preparation, maintenance, and organization of spatial data are rare during a disaster. In the chaos of the emergency response, good practices in spatial data management seem mundane and inconsequential—and are frequently compromised. Understandably, it seems more productive to create maps or conduct spatial analysis than to maintain your database or input metadata.

The deeper reality is that the reliability and utility of GIS—during and long after a humanitarian intervention—depend on how well the data is managed. Following good practices is not difficult and does not need to be onerous. ArcGIS offers uniquely powerful tools that allow you to organize and maintain your data efficiently so that you can concentrate on serving your customers.

This chapter shows you how to be disciplined with your spatial data management given the constraints of an emergency, ensuring your work brings maximum benefit to the people affected, both in the short- and long-term.

Spatial reference, coordinate systems, and map projections

GIS specialists in humanitarian organizations are faced with a unique problem: they never have enough time to prepare for the next assignment or recover from the last. If only they could arrive before a disaster, collect all the spatial data necessary to support emergency response, and organize it systematically. Then at the end of the mission, they would review, clean, and update it all so that the renewed data could be used to support long-term recovery and even the development of the region.

It doesn't happen that way. The field of humanitarian GIS remains relatively undisciplined in how spatial data is collected, shared, and maintained—a random set of shapefiles in a folder is generally what is used. Often very little, if any, information is provided on how the data was collected or by whom and when. Even some of the case-study data provided by organizations for use in this book falls short of the robust metadata standards advocated by the GIS industry.

Using pragmatic and efficient spatial data management skills during an emergency is key to winning the confidence of the decision makers you support and to ensuring that others can benefit from your efforts before, during, and after a disaster. The next few pages provide you with an overview of some core aspects of managing spatial data.

The **spatial reference** of a dataset is the series of parameters that define its coordinate system and other spatial properties. Each dataset is stored in the geodatabase using its spatial reference, so understanding spatial referencing systems is essential to your success with GIS. This is especially true during most humanitarian operations, when your data will be generated by multiple organizations using many different types of spatial reference, coordinate systems, and map projections.

Geographic data for any particular area is stored in separate layers. For example, roads will be stored as one layer, warehouses in another, and district boundaries in a third. To enable the data in each layer to integrate when displayed and queried, each layer must reference locations on the earth's surface in a common way. Coordinate systems provide this framework. They also provide the framework needed for data in different regions to be referenced in different ways. Each layer in the geodatabase has a coordinate system that defines how its locations are spatially referenced.

In the geodatabase, the coordinate system and other related spatial properties are defined as part of the spatial reference for each dataset. A spatial reference is the coordinate system used to store each feature class and raster dataset, along with other coordinate properties such as the resolution for x,y and optional z- and m- (measure) coordinates. If required, you can define a vertical coordinate system for datasets with z-coordinates that represent surface elevation.

ArcGIS includes spatial reference settings for the following:

- The coordinate system
- The precision with which coordinates are stored (coordinate resolution)
- Processing tolerances (such as cluster tolerance)
- The spatial or map extent covered by the dataset (spatial domain)

A **coordinate system** is used to represent the locations of geographic features, imagery, and observations (such as GPS locations) within a common geographic framework. Each coordinate system is defined by the following:

- Its measurement framework, which is either geographic or planimetric
- Units of measurement
- The definition of the map projection (for projected coordinate systems)
- Other measurement system properties such as a spheroid of reference, a datum, and projection parameters like one or more standard parallels, a central meridian, and possible shifts in the x- and y-directions

Two types of coordinate systems are used in GIS (see appendix A):

1. **Geographic coordinate systems (GCS)** are three-dimensional coordinate systems, such as latitude-longitude, typically expressed as degrees-minutes-seconds (DD:MM:SS) or decimal degrees (DD). The most common example is GCS_WGS_1984, although a variety of other spherical models are still in use in some parts of the world.
2. **Projected coordinate systems (PCS)** are two-dimensional (planimetric) coordinate systems that provide various methods to project the earth's surface onto a two-dimensional Cartesian plane, typically expressed in linear units such as meters or feet.

While a geographic coordinate system is essential when using data within a GIS, a projected coordinate system is an optional form of spatial reference selected according to the given map application. Consider these types of questions:

- Which spatial properties do you want to preserve?
- Where is the area you're mapping? Is your data in a polar region? An equatorial region?
- What shape is the area you're mapping? Is it square? Is it wider in the east–west direction?
- How big is the area you're mapping? On large-scale maps, such as street maps, distortion may be negligible because your map covers only a small part of the earth's surface. On small-scale maps (where a small distance on the map represents a considerable distance on the earth), distortion may have a bigger impact, especially if you use your map to compare or measure shape, area, or distance.

Asking these questions will enable you to select the best map projection model for your project. **Map projections** can be generally classified according to what spatial attribute they preserve:

- **Equal-area** projections preserve area. Many thematic maps use an equal-area projection. Maps of the United States commonly use the Albers equal-area conic projection.
- **Conformal** projections preserve shape and are useful for navigational charts and weather maps. Shape is preserved for small areas, but the shape of a large area, such as a continent, will be significantly distorted. The Lambert conformal conic and Mercator projections are common conformal projections.
- **Equidistant** projections preserve distances, but no projection can preserve distances from all points to all other points. Instead, distance can be held true from one point (or a few points) to all other points or along all meridians or parallels. If you will be using the map specifically to find features that are within a certain distance of other features, you should use an equidistant map projection.

- **Azimuthal** projections preserve direction from one point to all other points. This quality can be combined with equal-area, conformal, and equidistant projections, as in the Lambert equal-area azimuthal and the azimuthal equidistant projections.
- **Other projections** minimize overall distortion but don't preserve any of the four spatial properties of area, shape, distance, and direction. The Robinson projection, for example, is neither equal-area nor conformal but is aesthetically pleasing and useful for general mapping.

About projections and transformations

Maps are flat, but the surfaces they represent are curved. Transforming three-dimensional space into a two-dimensional map is called **projection.** Projection formulas are mathematical expressions that convert data from a geographical location (latitude and longitude) on a sphere or spheroid to a representative location on a flat surface. The process distorts at least one of these properties—shape, area, distance, direction—and often more. Because measurements of one or more of these distorted properties are often used to make decisions, anyone who uses maps as analytical tools should know which projections distort which properties, and to what extent. Briefly, conformal maps preserve shape; equal-area, or equivalent maps retain all areas at the same scale; equidistant maps maintain certain distances; and azimuthal or true-direction maps express certain accurate directions.

The Projections toolset contains tools to set a projection, reproject, or transform datasets. Properly projected data ensures accurate, reproducible representations and measurements with your GIS data, and is an essential data management skill for the humanitarian GIS professional.

Note: ArcGIS supports the following map projections:

ArcGIS Supported Map Projections

Projection	Type	Conformal	Equal Area	Equidistant*	True Direction*	Perspective	Compromise	Straight Rhumbs	World	Hemisphere	Continent/Ocean	Region/Sea	Medium Scale	Large Scale	North/South	East/West	Oblique	Equatorial	Midlatitude	Polar/Circular	Topographic	Geologic	Thematic	Presentation	Navigation	USGS
Aitoff	Modified Azimuthal	~	~				√		√										√					√		
Alaska Grid¹	Modified Planar	√	~		√							√							√				√			√
Alaska Series E	Pseudocylindrical											√									√					√
Albers equal area conic	Conic		√								√	√	√			√			√				√	√	√	√
Azimuthal equidistant	Planar			√	√				~	√	√	√		~			√	√	√	√	√				√	√
Behrmann equal area cylindrical	Cylindrical		√						√															√		
Berghaus Star	Interrupted, faceted			√		√	√		√								√							√		
Bipolar oblique conformal conic	Conic (Oblique)	√									√									√	√		√	√
Bonne	Pseudoconic		√								√				~										
Cassini-Soldner	Cylindrical												√	√						√					
Chamberlin Trimetric	Modified Planar			~							√														
Craster Parabolic	Pseudocylindrical		√						√																
Cube²	Faceted					√																			
Cylindrical equal area	Cylindrical		√	√										√		√							√		
Double Stereographic	Planar	√		√	√				√	√	√						√	√	√	√	√				√
Eckert I	Pseudocylindrical								√																
Eckert II	Pseudocylindrical		√						√																
Eckert III	Pseudocylindrical								√														√		
Eckert IV	Pseudocylindrical		√						√														√		
Eckert V	Pseudocylindrical								√														√		
Eckert VI	Pseudocylindrical		√						√														√		
Equidistant conic	Conic			√						~	√					√			√				√		
Equidistant cylindrical¹	Cylindrical			√									√												√
Fuller	Faceted					√			√														√	√	
Gall's Stereographic	Cylindrical						√		~															√	
Gauss-Kruger	Cylindrical (Transverse)	√									√	√	√	√	√		√	√			√	√			
Geocentric³	Spherical								√																
Geographic⁴	Spherical								√																
Gnomonic	Planar			√	√					~							√	√	√	√				√	√
Goode Homolosine⁵	Interrupted Pseudocylindrical Equal-Area		√						√														√	√	
Great Britain National Grid	Cylindrical	√									√	√	√								√	√			
Hammer-Aitoff	Modified Planar		√						√														√	√	
Hotine Oblique Mercator	Cylindrical (Oblique)	√									√	√	√	√			√			√				√
Krovak	Conic	√									√	√	√				√			√	√		√	√
Lambert Azimuthal equal area	Planar		√		√					√	√	√					√	√	√	√			√	√	√	√
Lambert conformal conic	Conic	√									√	√	√	√				√			√	√	√	√	√	√
Local Cartesian System	Planar												√												
Loximuthal	Pseudocylindrical						√																	√	
McBryde-Thomas Flat Polar Quartic	Pseudocylindrical		√						√																
Mercator	Cylindrical	√						√	~		√	√	√		√		√			√	√			√	√
Miller Cylindrical	Cylindrical						√		√														√		√
Mollweide	Pseudocylindrical		√						√														√		
New Zealand Grid	Modified Cylindrical	√										√	√			√			√	√			
Oblique Mercator	Cylindrical (Oblique)	√									√	√	√	√			√			√				√
Orthographic	Planar			√	√					√	√						√	√	√	√			√		
Perspective⁷				√	√					√	√						√	√	√	√			√		
Plate-Carée	Cylindrical			√																					√
Polar Stereographic	Planar	√		√	~				√	√	√								√	√		√		√
Polyconic	Conic			~			√					~	~							√				√
Quartic Authalic	Pseudocylindrical		√						√														√		
Robinson	Pseudocylindrical						√		√														√	√	
Rectified Skew Orthomorphic	Cylindrical (Oblique)	√									√	√				√			√				
Simple Conic	Conic			√						~	√				√			√				√		
Sinusoidal	Pseudocylindrical		√	~					√	√					√			√			√		√
Space Oblique Mercator	Modified Cylindrical	~											√			√			√				√
State Plane⁶		√										√							√	√	√	√		
Stereographic	Planar	√		√	√				√	√	√					√	√	√	√	√				√	√
Times	Pseudocylindrical						√		√														√	√	
Transverse Mercator	Cylindrical (Transverse)	√									√	√	√	√	√		√	√		√	√			√
Two Point Equidistant	Modified Planar			√					~		√	√					√					√	~	
Universal Polar Stereographic	Planar	√		√	~				√	√	√	√						√	√	√		√		√
Universal Transverse Mercator (UTM)	Cylindrical (Transverse)	√									√	√	√	√	√		√	√		√				√
Van der Grinten I	Circular						√		~														√	√	
Vertical Near-side Perspective⁷	Planar			√	√					√	√					√	√	√	√					
Winkel I	Pseudocylindrical								√														√		
Winkel II	Pseudocylindrical								√														√		
Winkel Tripel	Modified Planar						√		√														√	√	

1 Modified Stereographic Conformal
2 Used in ArcGlobe - true direction in limited areas.
3 Also known as Equirectangular
4 Not a map projection. The earth is modeled as a sphere or spheroid
5 Combination of the Mollweide and Sinusoidal projections
6 See Lambert Conformal Conic, Transverse Mercator, and Hotine Oblique Mercator
7 Also known as Perspective or Vertical Perspective

√ = Minimal Distortion
~ = Distortion is moderate for most of the area
* = Distortion is minimal in certain directions or at particular points

G-37891 9/09sp

Adapted from *Map Projections*, a USGS poster

Data management in ArcToolbox and ArcCatalog

ArcToolbox and ArcCatalog are the components of ArcGIS Desktop used to manage and organize geospatial data.

The ArcToolbox application provides an assortment of geoprocessing tools that can be categorized as (1) data analysis tools, (2) data conversion tools, and (3) data management tools. Depending on which version of ArcGIS is installed (ArcView, ArcEditor, or ArcInfo), you may have hundreds of tools bundled in "toolsets," which are contained in "toolboxes" within ArcToolbox.

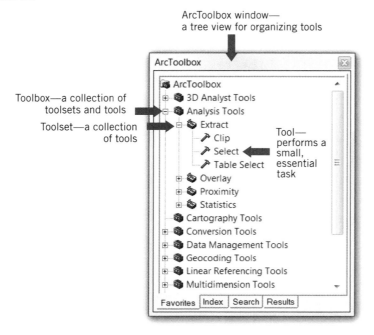

ArcToolbox's data analysis tools include the Analysis Tools toolbox, as well as dozens of other toolboxes that enable a wide range of analytical operations. In addition to some standard analytical tools, specialized tools are available as ArcGIS extensions. These optional software packages support the analytical needs of specialized users (military planners, business analysts, logisticians, statisticians, etc.), and are organized in separate toolboxes within ArcToolbox. Chapters 6-9 explore a few of these toolboxes to perform network analysis, spatial analysis, and statistical analysis.

Table 3.1 lists ArcToolbox's conversion and data management tools and their functions.

Toolset	Function
Conversion	
From Raster	Convert raster data to other formats
From WFS	Imports a feature type from a Web feature service (WFS) to a feature class in a geodatabase
Metadata	Validate metadata content according to a specific metadata standard or export metadata content to stand-alone metadata files that can be used with other metadata software
To CAD	Convert features to native CAD[a] format

Toolset	Function
To dBASE	Convert tables to dBASE format
To Geodatabase	Convert features and CAD files to geodatabase feature classes
To KML	Converts a feature or raster layer, or map document, into a Keyhole Markup Language (KML) file containing a translation of ESRI geometries and symbology
To Raster	Convert data to rasters
To Shapefile	Convert features to shapefiles
Data management	
Data Comparison	Compare one dataset with another and report any similarities and differences
Database	Increase database performance
Disconnected Editing	Check out or update data from a shared data system
Distributed Geodatabase	Replicate and dispense databases
Domains	Manage domains, both code and attribute, within a workspace
Feature Class	Perform basic feature class management functions, including copying and deleting features and appending multiple feature classes
Features	Transfer features from one feature type to another or attributes from one feature class to another
Fields	Make changes to the fields in the tables of a feature class
File Geodatabase	Compress and uncompress file geodatabase feature classes and tables
General	Perform simple dataset changes
Generalization	Simplify features within a feature class
Indexes	Add or remove spatial (and attribute) indexes from feature classes or relationship classes
Joins	Add or remove a table join
Layers and Table Views	Create and update layer files and table views
Projections and Transformations	Set the projection of and reproject a dataset[b]
Raster	Create and manipulate raster datasets
Relationship Classes	Create associations between feature classes and between feature classes and tables
Subtypes	Manage the subtypes of a feature class or a table
Table	Make adjustments to and query the table of a feature class
Topology	Manage the topological relationship among a number of feature classes
Versions	Make adjustments to versions of the data
Workspace	Create new workspaces and data
[a]CAD means computer-aided design. [b]See appendix A.	

Table 3.1   ArcToolbox conversion and data management tools

ArcCatalog helps GIS users by providing an integrated and unified view of all the data files, databases, and ArcGIS documents available to ArcGIS users. Unlike other data (such as a photo or a Word document), geographic datasets often consist of a set of files, rather than a single file. In Windows Explorer, a dataset appears as a list of system folders and files, whereas ArcCatalog displays and manages datasets as single entities.

ArcCatalog allows you to do the following (these and other descriptions of ArcGIS tools throughout the book are informed by ArcGIS Desktop Help):

- Connect to data source locations
- Browse and find geographic information
- Record, view, and manage data and associated metadata
- Define, export, and import geodatabase data models and datasets
- Search for and discover GIS data on local networks and the Web
- Create and manage the schemas of geodatabases

Geodatabases

Spatial data comes in many different formats, all of which can be managed in ArcCatalog and added to ArcMap as independent layers. These formats include coverages, shapefiles, Auto-CAD files, and geodatabases. Geodatabases are the most recent addition to the repertoire of spatial data formats.

Geodatabases organize data into feature classes. Much like its predecessor the shapefile, a feature class is a grouping of like geographic objects (such as roads or cities) that share the same attributes, spatial reference, and geometry type (points, lines, or polygons). For instance, the geodatabase Ghana.gdb might contain a polygon feature class of administrative units, a line feature class representing roads, another line feature class representing rivers, and a point feature class representing towns and villages.

Structurally, an ArcGIS geodatabase is a collection of geographic datasets held in a common file system folder, a Microsoft Access database, or a multiuser relational database (such as Oracle, Microsoft SQL Server, or IBM DB2). ArcGIS can handle databases of all sizes, from single-user databases built on files up to larger workgroup, department, and enterprise geodatabases accessed by many users.

A geodatabase may contain many dataset types, including the following:

- **Feature classes.** A feature class is a collection of geographic features that share the same geometry type (such as point, line, or polygon) and the same attribute fields for a common area. Streets, well points, parcels, soil types, and census tracts are examples of feature classes. Spatially related feature classes are often grouped together in a feature dataset.
- **Feature dataset.** A feature dataset is a collection of related feature classes that share a common coordinate system. Feature datasets are used to spatially or thematically integrate related feature classes. A transportation feature dataset may contain a feature class for highways, another feature class for secondary roads, a third feature class for unpaved or seasonal roads, and a point file for airports. The primary purpose of feature datasets is to organize related feature classes into a common dataset for building a topology, a network dataset, a terrain dataset, or a geometric network.
- **Tables.** The attributes and properties of geographic objects are stored and managed in tables. Tabular information is the basis of geographic features, allowing you to visualize, query, and analyze your data. All rows in an ArcGIS table have the same columns; each column stores a specific data type, such as a number, date, or character.
- **Raster datasets.** Raster datasets are individual rasters consisting of a single matrix of rows and columns. A mosaic of imagery would be considered a raster dataset.
- **Raster catalog.** A raster catalog is a collection of several raster datasets.

In addition to these dataset types, there are several other types of advanced datasets that can be stored within a geodatabase.

Metadata

Metadata is information that describes the content, quality, condition, origin, and other characteristics of data. It answers the following questions about GIS data:

- What is the subject matter?
- How, when, where, and by whom was the data collected?
- What is the data's availability and distribution protocol?
- What are the data's projection, scale, resolution, and accuracy?
- What is the data's reliability, and what instruments and methods were used to generate it?

Organization and storage

In ArcCatalog, metadata is divided into properties and documentation. Properties, such as the extent of a shapefile's features, are derived from the item by ArcCatalog and added to the metadata. Documentation is descriptive information (for example, legal information about using the resource) supplied by a person using a metadata editor tool. With the default settings in ArcCatalog, all you have to do to create metadata is click the item in the Catalog tree and then click the Metadata tab; properties will be added to the metadata automatically.

Each GIS resource has its own discrete metadata document. Metadata documents describing related resources are not interconnected. Metadata for a feature class describes only that feature class. It does not inherit any metadata from the feature dataset in which the feature class is stored.

At a minimum, metadata for a layer file should describe the layer file and the information it portrays. If a shapefile contains an entire suite of demographic data, its metadata should describe the values in each attribute column and its coordinate system since that is where those properties are defined and stored. Several layers may portray different aspects of the shapefile's data. One layer might show population growth, while another may show the ratio of retiree population to working population. The layer file's metadata should describe what the layer shows; how the data was classified, normalized, and symbolized; and any joins or relates that are defined in the layer—not all the details of the data.

Metadata created with ArcCatalog is stored in XML (Extensible Markup Language) format, either in a file alongside the item or within its geodatabase. In a geodatabase, metadata is stored in the GDB_UserMetadata table as a BLOB (binary large object) of XML data.

Standards

Following a well-known metadata standard is a good idea because tools already exist with which you can create and validate your metadata. If you plan to publish your metadata for a large audience, following a standard will also make it easier for people from different communities, industries, and countries to understand the information you provide because the standard acts as a dictionary, defining both the terminology and the expected values.

Several different metadata content standards have been established. The Federal Geographic Data Committee's (FGDC) Content Standard for Digital Geospatial Metadata (CSDGM) aims to provide a complete description of a data source. Because the CSDGM is quite detailed, other states and regions have created their own metadata standards to try to simplify the information that should be recorded. For example, the European Committee for Standardization (CEN), the Australia New Zealand Land Information Council (ANZLIC), and other organizations have created their own metadata standards and guidelines.

The International Organization for Standardization (ISO) technical committee TC211 develops standards for geographic information. ISO standard 19115, Geographic Information —Metadata, was designed for international use. It attempts to satisfy the requirements of all well-known metadata standards. ISO 19115 is a content standard, addressing what information should be included in metadata. It allows for detailed descriptions of geographic resources but has a small number of mandatory elements. ISO standard 19139, Geographic Information —Metadata—Implementation Specification, provides an XML schema that says how ISO 19115 metadata should be stored in XML format.

Many countries, regions, and communities are adopting ISO 19115 or ISO 19139, typically with some modifications and with their own XML schemas and document type definitions (DTD) specifying formatting requirements. Because many metadata standards and profiles exist, metadata in ArcGIS isn't required to meet any specific standard. However, you may be required to produce metadata that follows a standard.

Scenario: Creating a geodatabase for flood response in Ghana

In this chapter, you will simulate the process of building a geodatabase from various dataset types and a range of data sources. In doing so, you will learn many core data management skills, including ones you must employ throughout the remainder of this book, so take your time to really understand the techniques.

The table below lists the relevant attributes of each data layer used in this chapter. Some of this data was acquired in response to major flooding that occurred in 2007, in the West African country of Ghana. You will learn more about those floods in the next chapter. For the rest of this chapter, your objective is to acquaint yourself with the capabilities of ArcCatalog by building your first geodatabase.

Layer[a] or attribute	Description
Airports.shp	**Ghana major airports (points)**
NAME	Airport name
Cities.shp	**Ghana cities (points)**
CITY_NAME	City name
STATUS	National or provincial capital
POP_CLASS	Population classification
CommunitySurvey.shp[b]	**Ghana 2007 flood community survey (points)**
DamagedBridges.shp	**Ghana 2007 flood-induced bridge damage (points)**
BRIDGE_NAM	Bridge name
NMBR_DAMAG	Number of bridge segments damaged
Ghana_150m_EarthSat.img	**NaturalVue Landsat mosaic of Ghana**
Ghana.shp	**Ghana first administrative layer boundaries (polygons)**
Ghana_UTM30N.shp	**Ghana first administrative layer boundaries projected in UTM (polygons)**

Layer[a] or attribute	Description
HumanImpact.shp	**Ghana 2007 flood-induced human displacement and deaths/municipality (polygons)**
NAME	Municipality name
REGION	Region name
IDP	Number of internally displaced persons
DEATHS	Number of human deaths
InfrastructureDamage.shp	**Ghana 2007 flood-induced damage to houses, bridges, dams/ municipality (polygons)**
NAME	Municipality name
REGION	Region name
CLPSD_HSES	Number of collapsed houses
DES_BRG_CV	Destroyed bridges
DES_DAMS	Destroyed dams
Lakes.shp	**Ghana major lakes (polygons)**
AREA	Lake area
PERIMETER	Lake perimeter
LivestockLosses.shp	**Ghana 2007 flood-induced livestock deaths/municipality (polygons)**
NAME	Municipality name
REGION	Region name
DD_SWINE	Number of swine deaths
DD_CATTLE	Number of cattle deaths
DD_GOAT	Number of goat deaths
DD_SHEEP	Number of sheep deaths
DD_POULTRY	Number of poultry deaths
DD_DONKEY	Number of donkey deaths
MODIS_20070106_03.250m.jpg	**Ghana pre-2007 flood MODIS satellite image**
MODIS_20070106_04.250m.jpg	**Ghana pre-2007 flood MODIS satellite image**
MODIS_20070915_03.250m.jpg	**Ghana post-2007 flood MODIS satellite image**
MODIS_20070915_04.250m.jpg	**Ghana post-2007 flood MODIS satellite image**
PopulatedPlaces.shp	**Ghana gazetteer populated places (points)**
NAME	Populated place-name
Railways.shp	**Ghana major railways (lines)**
LENGTH	Railway segment length
Rivers.shp	**Ghana major rivers (lines)**
LENGTH	River segment length
Regions.shp	**Ghana second administrative layer boundaries (polygons)**
Roads.shp	**Ghana major roads (lines)**
LENGTH	Road segment length
UTM_Grid.shp	**World UTM grid polygons**
ZONE	Universal transverse Mercator coordinate system grid number
WestAfricaCountryOutlines.shp	**Ghana and surrounding countries (lines)**

[a]Layers that are not shapefiles (.shp) are raster datasets.
[b]Not populated to protect the identities of survey participants.

Table 3.2. Data dictionary

Exercise 3.1

Using ArcCatalog

So far in this book you have used ArcMap. You will now learn how to organize and manage various types of spatial and nonspatial data using ArcCatalog.

Reminder: A data dictionary appears in each chapter in this book. **Develop the habit of studying the data dictionary before beginning the exercises.** Likewise, whenever you receive a new dataset in your real work, always start by studying that dataset's origins and contents before using it. Only then can you use it effectively in your application, or share your outputs confidently with decision makers.

- Examine the chapter's data dictionary (table 3.2).
- Write down the names of a few thematic folders in which you might group the data layers (you will refer to this list later).

Launch ArcCatalog

1. **From the Windows taskbar, click Start > All Programs > ArcGIS > ArcCatalog.**

2. **Navigate to the ESRIPress > GISHUM folder.**

The left panel of ArcCatalog is called Catalog tree. It works much like Windows Explorer and allows you to open and close folders and to navigate data and storage locations.

3. **Expand the Chapter3 folder to see its contents.**

The right panel, called Catalog display, contains three tabs: Contents, Preview, and Metadata. When you choose the Contents tab, the datasets in the current folder are listed. The icon next to each file name indicates its type.

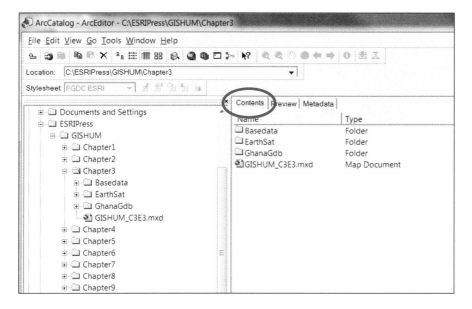

4. In the Catalog display, click the Contents tab (if it's not the default), then double-click the Basedata folder.

You will see 20 files: 16 shapefiles containing assorted point, line, and polygon data layers, and 4 raster datasets with MODIS satellite images.

5. Click WestAfricaCountryOutlines.shp, and then click the Preview tab at the top of the Catalog display.

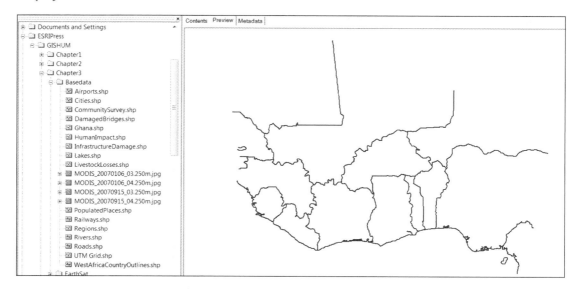

Previewing data this way allows you to get a quick glimpse of the data without loading it into a map. You can also use this tab to preview the contents of attribute tables.

6. At the bottom of the Catalog display, click the Preview drop-down arrow, and then select Table. Use the vertical scroll bar (or your mouse wheel) to view the various records contained in the layer.

7. Click the Metadata tab at the top of the Catalog display, and click and read the information on the Description, Spatial, and Attributes tabs (if these tabs are not visible, make sure the Stylesheet above the Catalog tree is set to FGDC ESRI).

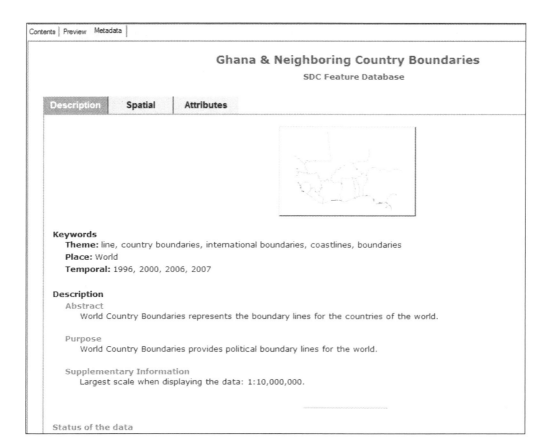

You'll learn more about metadata later in this chapter, but simply put, it is data about data. For example, the metadata for WestAfricaCountryOutlines.shp reveals that it is from the World Country Boundaries data that comes with ESRI Data & Maps; that it employs the WGS 1984 geographical coordinate system but no projected coordinate system; and that it can be displayed at a scale of up to 1:10,000,000.

You can also discover the layer's geoprocessing lineage (on the Spatial tab). For example, the original data was updated in March 2007 as follows:

Process step 2

Process description: The following steps were performed by ESRI: Added international boundary line between Serbia and Montenegro using the boundary from 2006 World Administrative Units. Removed an area in the Netherlands where land was in the process of being reclaimed from the sea but was never completed.

Process software and version: ArcView® GIS 3, ArcGIS® 9, ArcSDE®
Source used: ESRI Data & Maps 2006
Source used: CIA Factbook
Process date: 20070322

Your turn

Preview the geography and tables of each layer in the Chapter3 data folder and study each layer's metadata. Note the origin and completeness of each layer. You will learn how to improve the quality of metadata later in this chapter.

Exercise 3.2

Building a file geodatabase

Now that you are familiar with the exercise dataset, you are ready to create and populate a new file geodatabase devoted to flood response in Ghana.

Create a new file geodatabase

1.   Right-click the folder GhanaGdb in the Catalog tree, and then select New > File Geodatabase.

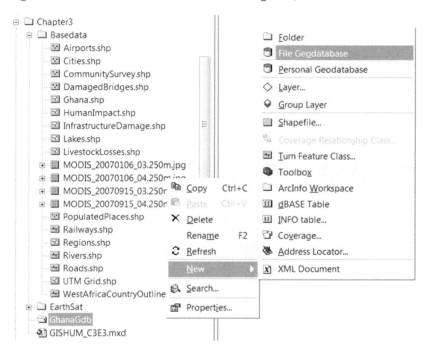

A new file geodatabase is created in the GhanaGdb folder.

2.   Name the file geodatabase **GhanaFlooding.gdb**.

Create a new feature dataset

1.   Click the GhanaGdb folder, making the GhanaFlooding.gdb database visible in the Catalog display.

2.   Right-click GhanaFlooding.gdb, and then select New > Feature Dataset.

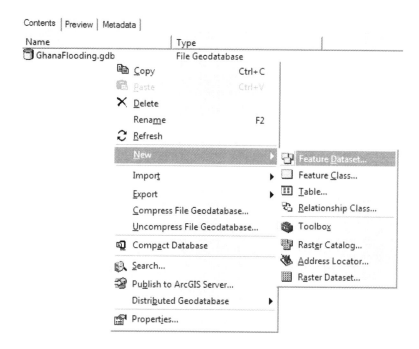

3. **Name the feature dataset Transportation.**

Perhaps Transportation was one of the thematic groups you proposed at the beginning of exercise 3.1. As described earlier, feature datasets are used to spatially or thematically integrate related feature classes that share the same spatial reference. Since we have three layers related to transportation (Airports.shp, Railways.shp, and Roads.shp), it makes sense to organize them in the same feature dataset.

4. **Click Next.**

You will now import the Airports, Railways, and Roads shapefiles as separate feature classes to the Transportation feature dataset. Since they must share the same coordinate system, import the spatial reference from the Cities.shp layer as the coordinate system for the Transportation feature dataset.

5. **Click Import.**

6. **Browse to the Basedata subfolder in your Chapter3 folder, click Cities, and click Add.**

The previous step assigns the coordinate system WGS 1984 to the feature dataset in order to ensure that the Transportation data layers have a consistent spatial reference to the Cities layer. You can also do this manually by selecting Geographic Coordinate Systems > World > WGS 1984, but since you know that the Cities layer contains the preferred spatial reference, you will just import it directly into your new feature dataset.

Note that WGS 1984 is a datum, as well as a spheroid, and can be associated with almost any projected coordinate system.

7. **Accept the default options for the next several screens by clicking Next, Next again, and then Finish.**

8. **Right-click the Transportation feature dataset, and then select Import > Feature Class (multiple).**

9. Browse to the Basedata folder, and then add the Airports, Railways, and Roads layers as input features.

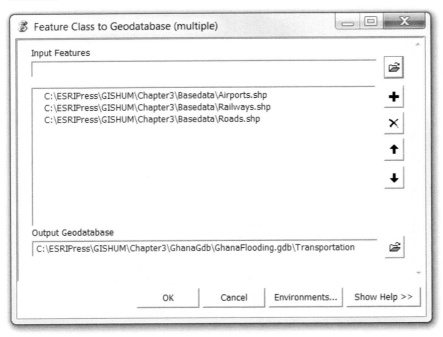

10. When your Feature Class to Geodatabase dialog box matches the preceding image, click OK.

11. When the import operation is complete, click Close.

Your GhanaFlooding geodatabase now has a Transportation feature dataset containing the three feature classes.

Your turn

Using the same method described, create feature datasets and load the other shapefiles into your geodatabase as follows. Use the same spatial reference (GCS_WGS_1984) for each feature dataset.

Administrative
        Cities.shp
        Ghana.shp
        PopulatedPlaces.shp
        Regions.shp
        UTM_Grid.shp
        WestAfricaCountryOutlines.shp

Your turn (continued)

DamageAssessment
        CommunitySurvey.shp
        DamagedBridges.shp
        HumanImpact.shp
        InfrastructureDamage.shp
        LivestockLosses.shp
Hydrology
        Lakes.shp
        Rivers.shp

ArcMap always tries to project all of the layers loaded into the table of contents using a consistent spatial reference; so even if the layers have different native coordinate systems, the first layer (base layer) loaded into an ArcMap session will determine the reprojection of all other layers loaded thereafter. By purposely assigning all layers of the GhanaFlooding geodatabase the spatial reference, you can avoid any complications that might occur if the layers need to be reprojected on the fly.

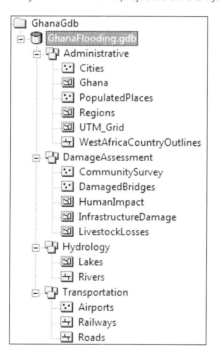

12. Before proceeding, ensure that your geodatabase matches the preceding image.

Note: The process previously described works for the exercise data because all the data is in GCS_WGS_1984. What do you do if your data is projected to a different GCS and datum?

Importing data to a feature dataset will project the data to the coordinate system of the feature dataset. However, it does *not* perform geographic (datum) transformations; so if the data being imported is on a different GCS, the data must be projected first, and the correct geographic transformation applied. Otherwise the imported data can be far away from the proper location with disastrous consequences for analysis and planning. A list of geographic transformations is available from the ESRI Support Center Web site: go to http://support.esri.com.

Create a new raster dataset

You are now ready to import the four MODIS satellite images and the one EarthSat satellite image from the Chapter3 folder to your geodatabase. Contiguous rasters can be stored as a single image or as a mosaic. If the rasters are not contiguous or are unrelated temporally or spatially, but have some other thematic similarity, they can be stored as a raster catalog (a related set of raster datasets).

1. **Right-click GhanaFlooding.gdb, and then select New > Raster Dataset.**

2. **Populate the Create Raster Dataset dialog box as follows:**

   - Raster Dataset Name with Extension: **MODIS_PreFlood**
   - Spatial Reference for Raster (optional): **GCS_WGS_1984**
   - Number of Bands: **3**

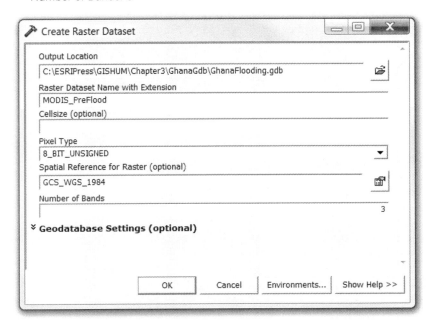

3. **Accept the default values for all other fields. When your settings match the preceding image, click OK.**

4. **When the operation is complete, click Close.**

The MODIS_PreFlood raster dataset should now appear in your geodatabase. You will now load both of the preflood MODIS images into this dataset so that they are mosaicked as a single image.

5. **Right-click the MODIS_PreFlood raster dataset, and then select Load > Load Data.**

6. **Add the two MODIS layers from January 6, 2007 (20070106), and click OK.**

7. **When the load operation is complete, click Close.**

The two images are now stored as a single raster dataset in your geodatabase.

8. Preview the new raster dataset in the Catalog display to confirm that the operation was successfully completed.

Your turn

Using the same method as above, create raster datasets and load the other images into your geodatabase as follows. Use the same spatial reference (GCS_WGS_1984) for each raster dataset.

MODIS_PostFlood
      MODIS_20070915_03.250m.jpg
      MODIS_20070915_04.250m.jpg

EarthSat_PreFlood
      Ghana_150m_EarthSat.img

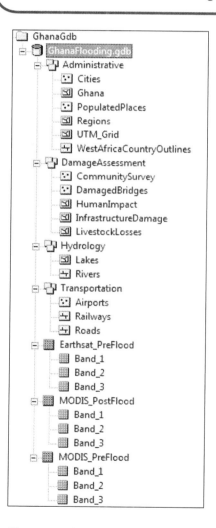

9. If your geodatabase operations were successful and your GhanaFlooding.gdb Catalog tree matches the preceding image, close ArcCatalog.

Exercise 3.3

Using coordinate systems

Within ArcGIS, every dataset has a coordinate system, which is used to integrate the dataset with other geographic data layers within a common coordinate framework such as a map. Coordinate systems enable you to integrate datasets within maps and to perform various integrated analytic operations, such as overlaying data layers from disparate sources and coordinate systems (see appendix A).

Understanding UTM

You'll often come across data that has been projected using the universal transverse Mercator (UTM) method. UTM is not a single map projection, but a system that employs a series of 60 zones, each of which is based on a specifically defined secant of the transverse Mercator projection coordinate system. The globe is divided into 60 north and south zones, each spanning 6 degrees of longitude. Each zone has its own central meridian. Zone 1 starts at 180° W and is followed by zones that travel 360 degrees eastward, ending with Zone 60.

Zone limits are 84° N and 80° S, with the equator separating north zones from south zones. The northern and southern polar regions use the universal polar stereographic coordinate system.

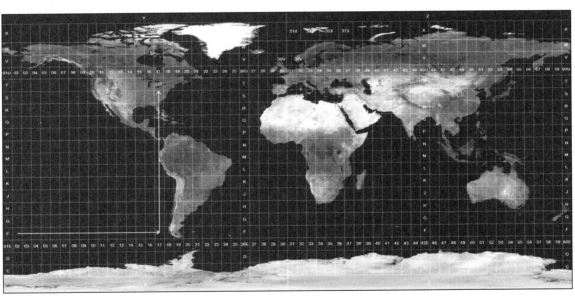

The UTM system is often incorrectly referred to as a projection, when in reality it is a collection of 60 separate projections, created to serve the mapping purposes of each of the 60 zones. The appeal of the UTM system is that it is globally recognized and suitable for mapping large parts of the earth, and often entire countries. It is a relatively reliable coordinate system for assigning projection information to new GIS files, especially files that will be shared with the wider geospatial community. The UTM coordinate system should never be used for data that extends more than 3 zones (18 degrees) east to west, due to the distortion introduced into the data by the limitations of the base projection, transverse Mercator. Ideally, the UTM coordinate system should be used only for data extending no more than 6 degrees east to west, the width of a single UTM zone.

1. **In ArcMap, open GISHUM_C3E3.mxd.**

A map of western Africa appears, showing the location of Ghana in relation to its neighbors. Overlaying the outlines of the countries is the UTM grid. The number in each cell indicates the UTM zone that applies to that part of the globe.

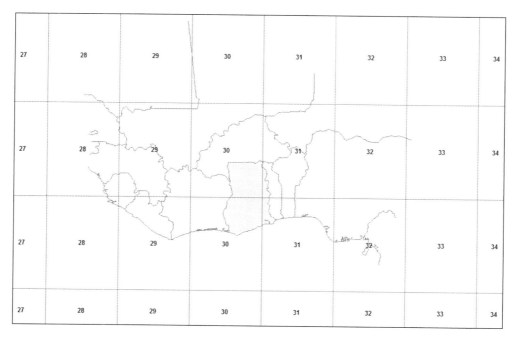

2. **On the main menu, go to Bookmarks and select Ghana.**

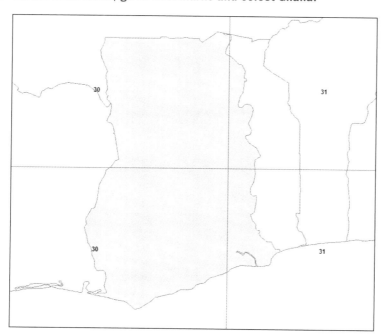

Notice that Ghana straddles two UTM zones, whereas Togo and Benin (the two countries to the east) are completely contained within a single UTM zone. You could, therefore, apply the UTM Zone 31 North (since the region is north of the equator) projection to Togo and Benin, but the situation for Ghana is more complicated.

UTM Zone 30 North covers most of Ghana, and it would be fine to use it if floods were affecting only the western half of the country. To analyze the entire country, we have two options:

1. Apply a single UTM projection (UTM Zone 30 North) to the whole country and sacrifice accuracy along the portion beyond that UTM zone's spatial extent.
2. Apply a zoneless projected coordinate system (such as Albers equal area conic, azimuthal equidistant Lambert conical) that can map the entire country uniformly, albeit more crudely.

Both options involve some margin of error, and only by manually comparing the two projections would you be able to know with certainty which is better.

Resolve unprojected or inconsistently projected data

If a data layer has a defined coordinate system, ArcMap can reproject it on the fly using the data frame's baseline coordinate system. You may at times encounter data that is missing a coordinate system or that is inexplicably rendered incorrectly in your map display (this is a frequent challenge during humanitarian emergencies). If a data layer does not have a defined coordinate system, ArcMap can display it, but it may not be spatially coincident with any other correctly projected layers in the data frame.

The next few steps will show you one method to resolve a common source of inconsistent map projection.

1.   **Start a new ArcMap session.**

Important: When attempting to resolve coordinate system issues, always open a new ArcMap session. This will ensure that your efforts aren't complicated with a preestablished default coordinate system.

2.   **Add the Ghana_UTM30N.shp layer to the map display.**

3.   **Add the CommunitySurvey.shp layer to the map display from the Basedata folder (not your GhanaFlooding geodatabase).**

You should now see the following warning screen:

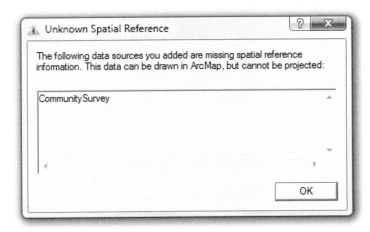

The warning screen reveals that the CommunitySurvey shapefile lacks spatial reference information.

4. Click OK.

What happens? Although the layer is added to the table of contents, it may not be visible.

5. Right-click the CommunitySurvey layer, and select Zoom To Layer.

You will now see the point features of the CommunitySurvey layer, but far away from where they should be located.

6. On the Tools toolbar, click Full Extent.

You should now be able to see both layers at a much smaller scale.

You should now see the entire country of Ghana, with one tiny dot to the east and south of the country.

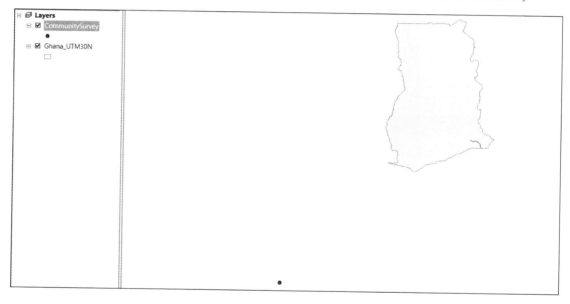

Using the Zoom tool, zoom repeatedly around this dot and you will realize that it is in fact the CommunitySurvey layer. As this layer lacks spatial reference information, it is unable to locate and resize correctly in relation to the correctly projected Ghana30N.shp layer.

Why is this so? Because the CommunitySurvey layer lacks spatial reference, it cannot "recognize" the spatial reference of the second layer Ghana30N.shp, and therefore the CommunitySurvey layer cannot be projected on the fly to align with Ghana_UTM30N.shp.

Unfortunately, no single solution can correct all spatial reference problems, but you can follow a number of key steps to determine the nature of the error. (See *Lining Up Data in ArcGIS: A Guide to Map Projections* by Margaret M. Maher [ESRI Press, 2010]). Oftentimes, datasets that exhibit the problem of a missing coordinate system are derived from GPS units that failed to record the exact coordinate system (possible reasons include human error or unit malfunction; coordinate information lost in transit through miscopied files or undelivered e-mail attachments). It's safe to assume that this information was collected by GPS, and if so, then reverting to the default coordinate system used by GPS will correct the issue. The default coordinate system used by the GPS system is the geographical coordinate system WGS 1984.

7. Right-click the word Layers at the top of the table of contents and scroll to Properties. (Another way to access the Data Frame Properties dialog box is to right-click with your mouse pointer sitting on the map itself. Try it!)

8. In the Data Frame Properties dialog box, click the Coordinate System tab.

9. Observe the current coordinate system settings, and then navigate to Predefined > Geographic Coordinate Systems > World > WGS 1984 within the Select a coordinate system box.

10. When your settings appear as follows, click OK.

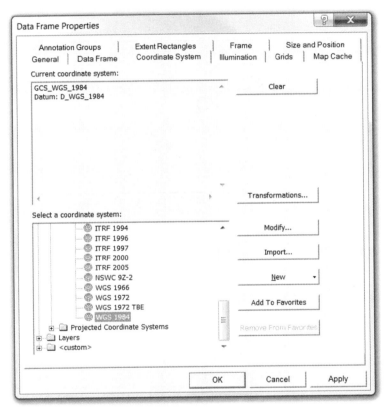

11. Your point data should now align correctly with your other data layers for Ghana. (If the objects in your map disappear, click once again on the Zoom to full Extent button)

This experiment has shown that, as suspected, the data was indeed unreferenced GPS data (which is often the case). Now you can define the geographic reference system (or spherical reference system) for this file. You will reproject it to a projected coordinate system more appropriate for Ghana.

12. On the Standard toolbar, click the ArcToolbox 🔲 button.

The ArcToolbox menu now appears on your screen. You will use the Projections and Transformations toolset now (and learn about other toolsets in later chapters).

13. Expand the Data Management Tools toolbox, and then go to Projections and Transformations > Define Projection.

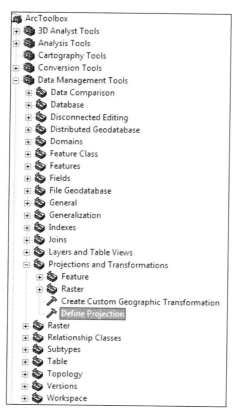

14. In the Define Projection tool wizard, select the CommunitySurvey shapefile as the Input Dataset or Feature Class.

The tool wizard indicates that the layer has an unknown coordinate system.

15. Click the Spatial Reference Properties button next to the Coordinate System field.

16. In the dialog box that pops up, click Select, go to Geographic Coordinate Systems > World > WGS 1984.prj, and then click Add and OK.

17. When your settings appear as follows, click OK.

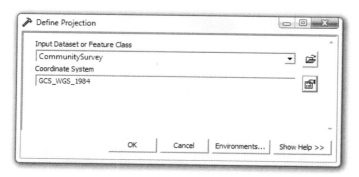

18. **Add the newly projected CommunitySurvey shapefile from the Basedata folder.**

If your community survey data was not correctly projected within the boundaries of Ghana originally, it is now, since our guess that the missing spatial reference was GCS WGS 1984 proved correct.

Important: If the CommunitySurvey layer had been opened with a layer that shared its correct (but missing) coordinate system information (such as the original Ghana.shp layer), then it would have immediately drawn in the correct location. In the example above, you were able to resolve a misprojection because you correctly guessed the original coordinate system used to collect the CommunitySurvey layer.

This exercise introduced you to some simple steps you can use to investigate and correct GIS data lacking a defined coordinate system.

Tip: Always make a note of the coordinate system and methods used to collect any data you receive or gather yourself, and add that to your metadata.

Exercise 3.4

Managing metadata

Metadata is an information file that describes the basic characteristics of a data resource. Usually presented as an XML document, it conveys the who, what, when, where, why, and how of that information.

A geospatial metadata record includes core library catalog elements, such as title, abstract, and publication data; geographic elements, such as geographic extent and projection information; and database elements, such as attribute label definitions and attribute domain values (from `http://www.fgdc.gov/metadata`).

Any item in ArcCatalog, including folders and file types such as Word documents, may contain metadata. Metadata is copied, moved, and deleted along with the item when it is managed with ArcCatalog.

Import metadata

You have one final operation before your GhanaFlooding geodatabase is complete. The EarthSat image has an XML metadata file that needs to be manually imported into the EarthSat raster dataset.

1. **In ArcCatalog, click the EarthSat_PreFlood raster dataset in the Catalog tree and view its metadata.**

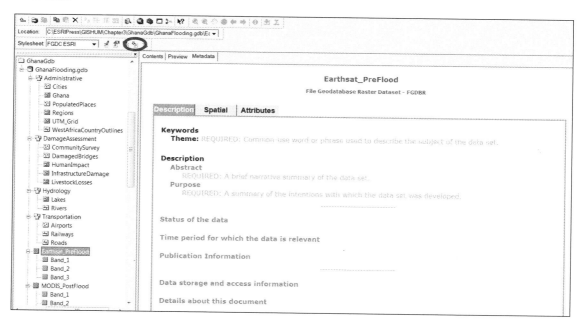

2. **Click the Metadata tab, and then click the Import metadata** 📥 **button on the Metadata toolbar.**

Metadata can be imported in various formats, including Standard Generalized Markup Language (SGML), Extensible Markup Language (XML), and text (TXT).

3. **Change the default SGML format choice to XML, and then click Browse. Navigate to the EarthSat folder in the Chapter3 data folder.**

The metadata file is in XML format, indicated by the icon preceding the file name.

4. **Select Ghana_150m_EarthSat.img, and then click Open.**

5. **Click OK in the Import Metadata dialog box.**

The metadata documentation and properties have now been added to the EarthSat raster dataset in your geodatabase.

Create an image thumbnail

A small preview image, called a "thumbnail," is missing from the top of the EarthSat_PreFlood metadata page.

1. **Click the Preview tab at the top of the Catalog Display.**

2. **With the preview visible, click the Create Thumbnail ⊞ button on the Standard toolbar.**

Return to the EarthSat_PreFlood raster dataset's metadata. You should now see the image thumbnail. Whenever making significant changes to any data layer, use this simple process to refresh the image thumbnail in that layer's metadata.

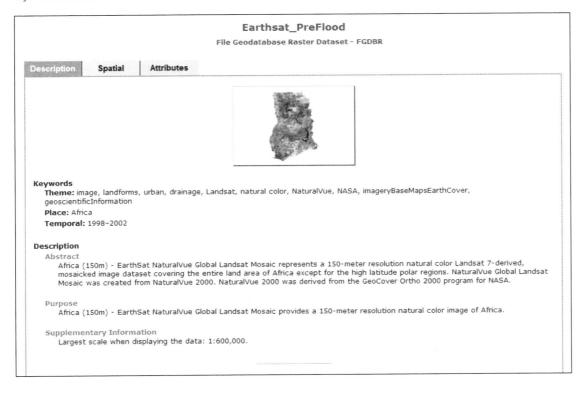

Your metadata is now populated with all of the original information that was provided by Earth Satellite Corporation and ESRI, the original providers of the NaturalVue Global Landsat Mosaic.

View metadata properties

The EarthSat imagery dataset, provided with ESRI Data & Maps, offers a good example of a well-developed set of metadata. You will now find out how ESRI's EarthSat imagery was created, which individuals and organizations are associated with the data, what geoprocessing has been performed on the data, and other important facts about the dataset.

1. **Read the metadata associated with the dataset under the Description and Spatial tabs.**

Since no attribute data is associated with a raster dataset, no metadata is listed under the Attributes tab.

2. **Click the green paragraph headers to open and close various elements of metadata.**

By default, metadata is presented in the format of the FGDC ESRI stylesheet. Spatial metadata is automatically updated by ArcCatalog; however, description metadata must be manually updated by the user.

3. **Click the Stylesheet drop-down arrow on the Metadata toolbar, select FGDC FAQ, and peruse the frequently asked questions.**

4. **Click the Stylesheet drop-down arrow again, and then select FGDC.**

As you can see, various stylesheets present different amounts of metadata in different ways.

Earthsat_PreFlood

Data format: File Geodatabase Raster Dataset

File or table name: Earthsat_PreFlood

Coordinate system: Geographic

Theme keywords: image, landforms, urban, drainage, Landsat, natural color, NaturalVue, NASA, imageryBaseMapsEarthCover, geoscientificInformation

Abstract: Africa (150m) - EarthSat NaturalVue Global Landsat Mosaic represents a 150-meter resolution natural color Landsat 7-derived, mosaicked image dataset covering the entire land area of Africa except for the high latitude polar regions. NaturalVue Global Landsat Mosaic was created from NaturalVue 2000. NaturalVue 2000 was derived from the GeoCover Ortho 2000 program for NASA.

FGDC and ESRI Metadata:

- Identification Information
- Data Quality Information
- Spatial Data Organization Information
- Spatial Reference Information
- Entity and Attribute Information
- Distribution Information
- Metadata Reference Information
- Geoprocessing History
- Binary Enclosures

Metadata elements shown with blue text are defined in the Federal Geographic
Elements shown with green text are defined in the *ESRI Profile of the CSDGM.* El
ArcCatalog adds hints indicating which FGDC elements are mandatory: these a

Earthsat_PreFlood

Frequently-asked questions:

- What does this data set describe?
  1. How should this data set be cited?
  2. What geographic area does the data set cover?
  3. What does it look like?
  4. Does the data set describe conditions during a particular time period?
  5. What is the general form of this data set?
  6. How does the data set represent geographic features?
  7. How does the data set describe geographic features?
- Who produced the data set?
  1. Who are the originators of the data set?
  2. Who also contributed to the data set?
  3. To whom should users address questions about the data?
- Why was the data set created?
- How was the data set created?
  1. Where did the data come from?
  2. What changes have been made?
- How reliable are the data; what problems remain in the data set?
  1. How well have the observations been checked?
  2. How accurate are the geographic locations?
  3. How accurate are the heights or depths?
  4. Where are the gaps in the data? What is missing?
  5. How consistent are the relationships among the data, including topology?
- How can someone get a copy of the data set?
  1. Are there legal restrictions on access or use of the data?
  2. Who distributes the data?
  3. What's the catalog number I need to order this data set?
  4. What legal disclaimers am I supposed to read?
  5. How can I download or order the data?
  6. Is there some other way to get the data?
  7. What hardware or software do I need in order to use the data set?

5. **Return to the FGDC ESRI metadata stylesheet.**

6. **Click the Spatial tab under Metadata. Underneath the name of the coordinate system used by the data, click Details to reveal the EarthSat_PreFlood raster dataset's spatial reference.**

7. **Scroll down to see the dataset's extent under Bounding coordinates.**

The bounding coordinates, which describe the spatial extent of the layer, are presented in decimal degrees according to the coordinate systems in use. As we are using only WGS 1984, the geographic and projected x,y coordinates are identical. If UTM 30 North were also in use, however, projected (local) coordinates would differ from the geographic coordinates.

8. **Scroll farther down to see the dataset's lineage.**

Lineage
  FGDC lineage
   Process step 1
      Process description: EarthSat NaturalVue Global Landsat Mosaic is a 150-meter resolution natural color Landsat 7-derived, mosaicked image dataset covering the entire land area of the Earth except for the high latitude polar regions. NaturalVue Global Landsat Mosaic was created from NaturalVue 2000™. NaturalVue 2000 was derived from the GeoCover Ortho 2000 program for NASA. The NASA mosaics are band combination 7,4,2 which is excellent for geology, but which results in vivid green vegetation, and purple cities and snow. As the name implies, NaturalVue 2000 has more natural colors acceptable for a wider viewing audience. NaturalVue 2000 is derived from imagery from the year 2000 +/- 2 years. An extensive search was conducted to get the most cloud-free imagery.
      Source produced: EarthSat - NaturalVue
      Process date: 200311

   Process step 2
      Process description: The following steps were performed by ESRI: Assembled the GeoTIFF files provided by EarthSat into continental units. Processed these files with the JPEG 2000 encoder available from LizardTech™ reducing the size from 55 gigabytes to less than 6 gigabytes.
      Source used: EarthSat - NaturalVue
      Process date: 20040227

   Process step 3
      Process description: The following steps were performed by Firoz Verjee: Clipped the continental GeoTIFF files provided by ESRI using the Ghana country polygon, and then imported the associated metadata.
      Source used: Chapter 3 – Earthsat - NaturalVue
      Process date: 20090929

   Process step 4
   Process step 5
   Process step 6
   Sources
      Source 1: NaturalVue Global Landsat Mosaic (EarthSat - NaturalVue)

The lineage metadata, generated in part manually and in part automatically, reveals that the dataset was produced from a mostly cloud-free mosaic of Landsat 7 band combination 7, 4, 2 data collected over the period 1998–2002.

You can also see that on February 27, 2004, ESRI then used a JPEG2000 encoder to reduce the original file size of the GeoTIFF imagery provided by EarthSat.

Next, you notice that on September 29, 2009, the author of this book clipped the original ESRI data, which covered all of Africa, using a polygon file to limit the coverage to Ghana's national boundaries.

Additional processing steps will reveal other operations performed since then, including the operations you performed earlier in this exercise.

Create and update metadata

The EarthSat_PreFlood metadata is very robust; however, the layer's place and its purpose need to be updated to reflect that the image depicts only Ghana, not all of Africa.

1. On the Metadata toolbar, click the Edit metadata 📝 button.

2. In the Editing dialog box, add **Ghana** to the beginning of the Abstract and Purpose fields to clarify the true coverage of the layer.

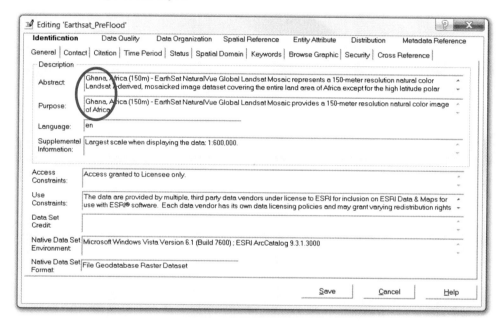

3. Click through the other tabs and make additional improvements to the metadata.

4. When you are satisfied that the EarthSat_PreFlood layer's metadata is as complete as possible, click Save.

Now that you are familiar with FGDC standard metadata, you will improve the metadata of another layer, which is comparatively incomplete.

5. View the metadata of the CommunitySurvey layer in the DamageAssessment feature dataset of your geodatabase.

The metadata's description page has not been populated. Using the information contained in the author's e-mail on the next page, you will attempt to improve the CommunitySurvey layer's metadata.

6. Click the Edit Metadata button to begin your process of improving the documentation.

The information in an FGDC metadata document is divided into seven main categories:

1. Identification
2. Data Quality
3. Spatial Data Organization
4. Spatial Reference
5. Entity and Attribute
6. Distribution
7. Metadata Reference

The primary tabs across the top of the FGDC metadata editor box correspond to these categories. When you click a primary tab in the editor box, secondary tabs will help you edit specific subcategories of metadata.

At first, the organization of information in the FGDC editor box may seem overwhelming, but as you become familiar with the interface, you will find it easier to locate the elements that you want to edit. The Help button in the bottom right-hand corner of the editor box is an excellent resource to guide you through populating the various windows. Most important, it shows you where to find the mandatory FGDC elements.

Elements that are mandatory in the FGDC standard appear with red text in the editor box. If you complete this suggested documentation, your metadata will satisfy the FGDC standard's minimum requirements.

For most elements, text boxes are provided in which you can type the appropriate information. When the standard expects one of several predefined values, a drop-down list lets you choose among them.

7. Click the Identification tab (if not already selected).

8. Using information from the author's e-mail, go to the General tab and complete the fields as well as you can.

9. Go the Contact tab and click Details. Add contact information to complete the available fields as well as you can, based upon your understanding of who should be contacted regarding the data.

10. Go the Citation tab, and then click Details. Base your citation information on the bibliographical reference for this book by ESRI Press.

11. Go to the Time Period tab, and then indicate that the dataset was developed between September 1 and September 30, 2007.

12. Go to the Status tab, and then indicate that the dataset is complete and that no further updates are planned.

13. Go to the Keywords tab, and then enter some appropriate keywords into the various boxes. For example, you could type **Community Survey** into the Theme Keyword box.

The metadata editor tool allows you to add multiple keywords to the metadata file, so you may also wish to add **Census** as a theme keyword. Use the Help function if you need extra guidance or clarification at any time.

14. Go to the Security tab, and if your organization normally requires you to do so, rate the CommunitySurvey layer as Unclassified.

You have now completed the Identification metadata fields. These fields are the minimum requirements for any layer; however, you can still populate other categories of metadata with the knowledge you have about the CommunitySurvey layer.

15. Using the same process as above, go to each category and subcategory tab of the layer's metadata and do your best to improve upon the current records.

16. Click Save after you finish, and examine how your work appears in different stylesheets.

Your turn

Using the methods learned in this chapter, review each of the data layers in your geodatabase and improve their metadata. You may need to research the 2007 floods that occurred in Ghana. You will need to learn about how MODIS imagery is collected, since neither the preflood nor the postflood images in your exercise folder include any metadata.

What to turn in

If you are working in a classroom setting with an instructor, submit an electronic or printed version of the description metadata for each data layer in your geodatabase. Remember to click the green headers to expand each section of the Description page before going to File > Print Metadata.

You have finished building the GhanaFlooding geodatabase. If this was your first opportunity to create a geodatabase and learn about metadata, don't worry if you feel a little overwhelmed. Other chapters in this tutorial reinforce many of the skills introduced so far, and with additional practice, you will feel comfortable in managing spatial data effectively.

Spend some time going over what you've learned so far. You may also want to examine ArcGIS Help files related to geodatabases, map projections, and metadata creation. ESRI provides extensive reference materials at `http://support.esri.com/` and in its desktop software.

The humanitarian community needs GIS professionals with good data management habits. Because of the lack of time and the pressure to produce tangible outputs, specialists in humanitarian GIS have had a tendency to compromise proper spatial data management practices. That's why it's important to exercise these skills until they become almost second nature to you. Effectively managing your organization's spatial data will benefit you, your organization, and the humanitarian community at large.

OBJECTIVES

Discover public sources of spatial data
Digitize satellite imagery into feature data
Import and reformat x,y coordinate data
Export to and view data on ArcGIS Explorer and Google Earth

Chapter 4

Generating spatial data during emergencies

Ask anyone familiar with the use of GIS during disasters and they will undoubtedly complain about the lack of spatial data. Humanitarian emergencies commonly occur in regions with poor spatial data infrastructure and weak data sharing mechanisms to start with. Even the spatial data you do have can quickly become outdated in the chaotic, dynamic conditions of disaster. Unless new data is periodically collected and added to your geodatabase, you will be limited in your ability to support better decision making.

Understanding how to populate a geodatabase is therefore an essential skill for the humanitarian GIS specialist. This chapter explores some of the sources and methods with which to create, find, integrate, and share spatial data using ArcGIS.

Populating a geodatabase

Spatial data can be generated before, during, and after humanitarian emergencies in a variety of ways. Private firms, geospatial consortiums, the United Nations, and host governments are common providers of raster, vector, tabular, and other types of data for GIS users. In the United States, federal government agencies and many local governments provide public access to high-quality spatial data either through data portals on their Web sites or upon request.

In countries that lack a spatial data infrastructure or do not make their spatial data publically available, it can be a little more challenging for a humanitarian organization to populate its geodatabase. GIS specialists must respect the security concerns and regulations of a host country. There is rarely enough time to realize full potential by any cartographic or analytical measure. GIS specialists need to be efficient and resourceful in gathering data and pragmatic in how they use it.

Humanitarian organizations commonly gather spatial data using the following resources:

1. **ESRI World Data & Maps.** Several DVDs of spatial data are included with the purchase of any ArcGIS software license:
   a. Basemap and thematic ArcMap documents (MXDs) for Canada, Europe, Mexico, the United States, and the World
   b. Commercial raster and vector data from several private and nonprofit organizations
   c. 90-meter Shuttle Radar Topography Mission (SRTM) global elevation data
   d. A rich set of spatially referenced national census data and basic streetmap data for the United States
2. **Geographic data portals.** Anyone with high-speed Internet access can search, download, and publish spatial data using various data portals, such as these:
   a. ArcGIS Online: http://www.arcgis.com
   b. US Government Geospatial One Stop: http://www.geodata.gov
   c. UN Spatial Data Infrastructure: http://www.geonetwork.nl/
   d. UN Geographic Information Working Group (multiple links to UN data portals): http://www.ungiwg.org/data.htm
   e. Columbia University's CIESEN: http://www.ciesin.org/data.html
   f. Stanford University's GIS Department: http://www-sul.stanford.edu/depts/gis/
   g. Famine Early Warning Systems Network (FEWS NET), Africa Data Dissemination Service: http://earlywarning.usgs.gov/adds/
3. **Global Spatial Data Infrastructure Association.** Many countries and organizations have begun to invest in the maintenance of national or regional spatial data infrastructure (SDI). To learn more about this trend and to search for SDI links suitable for your areas of interest, visit the Global Spatial Data Infrastructure Association Web site: http://gsdi.org/SDILinks.asp
4. **Local organizations.** Establishing data-sharing relationships with the governments and institutions in the regions where you are working can be one of the most effective—and mutually-beneficial—strategies for generating spatial data.
5. **Manual data collection.** Global Positioning System (GPS) receivers have become an indispensible tool for mapping, analyzing, storing, and communicating features of interest. The availability of precision location information (latitude, longitude, and elevation) to anyone with a low-cost receiver, anywhere in the world, is one of the key reasons for the growth of GIS usage since the mid-1990s. You will learn how to import GPS data into your geodatabase later in this lesson.

6. **Data association.** Tables that contain village names, postal codes, and population can be joined to point or polygon spatial data layers, with the join based on a common field. Such an example of data association is the UN's system of P-Codes (place codes). The UN Office for the Coordination of Humanitarian Affairs is responsible for producing P-Codes and for serving as an information clearinghouse during major humanitarian emergencies. Study its activities around the world at http://ochaonline.un.org/Geographic/tabid/1084/Default.aspx

Scenario: Major flooding in West Africa

In the last chapter you built a file geodatabase using data related to Ghana. To review, in 2007 West Africa experienced devastating seasonal floods that killed and displaced large numbers of people and destroyed significant amounts of transportation infrastructure, food stocks, and livestock.

To determine how much aid was required, the UN sent its Disaster Assessment and Coordination (UNDAC) team to Ghana. UNDAC teams are designed to travel rapidly to disaster-stricken countries to estimate the humanitarian impact of a disaster. In addition to the UNDAC team, a nongovernmental organization that specializes in GIS-based humanitarian services, MapAction UK, was invited to assist with data collection and production of situation maps. (MapAction provides mapping support during the immediate aftermath of humanitarian emergencies and relies primarily upon volunteers to undertake its missions around the world.)

MapAction's team leader and logistician deployed within 24 hours of receiving the UN's request. Once in the flood-impacted areas, MapAction needed to produce orientation maps and situation reports immediately, using whatever base data and additional data they could generate. Knowing that, the rest of the team also moved quickly to procure as much relevant geospatial data as they could prior to departing the home base in England.

In this lesson you will perform some of the data generation tasks that the team from MapAction needed to do before and after their arrival in Ghana.

Layer or attribute	Description
AffectedCommunities.dbf	**Ghana affected communities coordinates (table)**
FULL_NAME	Full name of affected community
DISTRICT	District name
D_LAT	Latitude (degrees)
M_LAT	Latitude (minutes)
S_LAT	Latitude (seconds)
D_LONG	Longitude (degrees)
M_LONG	Longitude (minutes)
S_LONG	Longitude (seconds)
DamagedSchools.txt	**Damaged schools coordinates (comma-separated values)**
SCH_NAME	Schools name
COMMUNITY	Community name
DISTRICT	District name
DD_LAT	Latitude (decimal degrees)
DD_LONG	Longitude (decimal degrees)

Table 4.1 Data dictionary

Exercise 4.1

Gathering spatial data using geodata portals

Becoming aware of immediate needs that could have far-reaching effects can be an effective mindset for the initial stage of gathering base data. For your source of baseline data in this exercise, a portal called the FEWS NET Africa Data Dissemination Service (ADDS) is particularly appropriate. FEWS NET stands for the U.S. Agency for International Development's Famine Early Warning Systems Network. Its information system is designed to identify problems in the food supply system that potentially could lead to famine or other food-insecure conditions in sub-Saharan Africa, Afghanistan, Central America, and Haiti.

FEWS NET is a multidisciplinary project that collects, analyzes, and distributes regional, national, and subnational information to decision makers about potential or current famine and hazards related to climate or socioeconomic situations. With such information, they can authorize timely measures to prevent food-insecure conditions in these nations. Regions and countries with FEWS NET representatives include sub-Saharan Africa (Angola, Burkina Faso, Chad, Djibouti, Ethiopia, Kenya, Malawi, Mali, Mauritania, Mozambique, Niger, Nigeria, Rwanda, Somalia, southern Sudan, Tanzania, Uganda, Zambia, and Zimbabwe), Central America (Guatemala, Honduras, and Nicaragua), Afghanistan, and Haiti (see `http://earlywarning.usgs.gov/adds/overview.php`).

Download geodata

1. **Open your Internet browser, and then navigate to the following Web site:** `http://earlywarning.usgs.gov/adds/`.

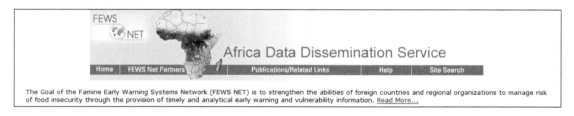

The Goal of the Famine Early Warning Systems Network (FEWS NET) is to strengthen the abilities of foreign countries and regional organizations to manage risk of food insecurity through the provision of timely and analytical early warning and vulnerability information. Read More...

2. **Scroll down until you come to the section entitled "Data Selection for download."**

3. **In the By Theme box, click the radio button for Digital Map Data.**

The next page lists the various spatial data categories available. Notice that Ghana is listed under two categories only, Administrative Boundaries and West African Spatial Analysis Project.

4. First, click the Administrative Boundaries hyperlink. You will see the following interface:

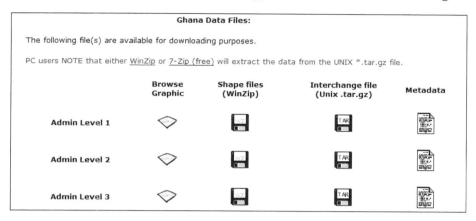

5. Examine the browse graphic and metadata of each data file.

Consider how the Admin Level 1, 2, and 3 files might complement your GhanaFlooding geodatabase from chapter 3. The feature classes of Level 1 (Ghana) and Level 2 (Regions) are loaded into your Administrative feature dataset, but you don't have Level 3 administrative boundaries. It is not possible to know from the metadata if the quality of the ADDS data is better or worse than what is currently in your geodatabase.

6. **If your computer does not already have WinZip or another similar software package, download and install such software using the links provided on the same Web page. (You may need to obtain authorization from your system administrator to install new software onto your computer.)**

How to configure WinZip to avoid data corruption

Warning: WinZip will corrupt data if used with default settings!

To avoid data corruption, follow these steps, which apply to all versions of WinZip, to configure the program:

1. Open WinZip Classic.
2. Select Options > Configuration > Miscellaneous tab.
3. Clear the "TAR file smart CR/LF conversion" check box. The box remains empty until it is checked again.
4. Unzip and import the data.
5. Because the files were compressed using UNIX into .TAR.GZ, the following steps will avoid corrupting these data during the decompression process:
   a. When the .TAR.GZ file is double-clicked to open, WinZip opens a dialog box that reads: "Archive contains X number of files.
       <file_name(s).tar>
       'Should WinZip decompress it to temporary folder and open it?'"

   Click NO.

   b. Check the "TAR file smart CR/LF conversion" box. Extract the TAR file from the .GZ archive to the desired folder.
   c. Double-click the TAR file to start WinZip a second time. Clear the "TAR file smart CR/LF conversion" check box, and extract the TAR file.

7. **Download the Admin Level 3 file, go to the previous Web page, and click the WinZip 💾 icon under Shapefiles (WinZip).**

8. **Save and extract the data to the FEWSNET folder located in the Chapter4 folder. By default, the new file is called ghadmin3.shp.**

Compare the Level 1 and Level 2 data with the data already loaded into your geodatabase, just in case it is superior.

9. **Download the remaining administrative files to the same folder.**

10. Examine the West African Spatial Analysis Project file's metadata.

Aside from being in a unique format (Unix TAR, or tape archive, format), the metadata reveals that the file offers only minor additional value to your existing geodatabase and is based upon information collected more than 15 years ago. Nonetheless, you may wish to download the file and see that data.

Let's examine the downloaded Admin Level files using ArcMap.

11. Open ArcMap, and then add the ADDS files to your table of contents.

12. Symbolize the map appropriately, so that you can see the geographical hierarchy. Your map should look something like the one that follows:

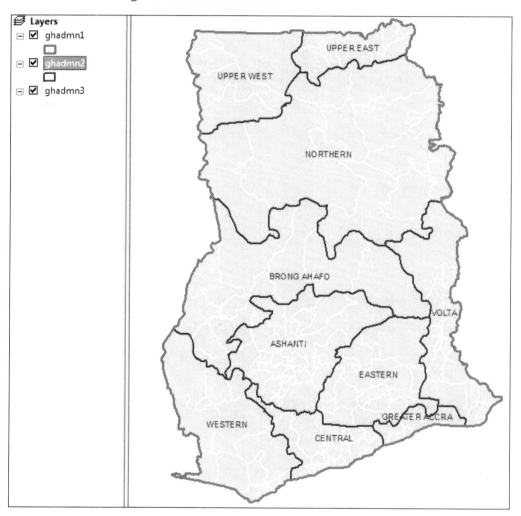

Use the following settings if you want to replicate the symbology in the preceding map:

- ghadmin1.shp—Use a Gray 70 percent outline of width 2.0 and no fill color.
- ghandmin2.shp—Use a Tuscan Red outline of width 1.5 and no fill color. Also, add the district names as labels using the same color (go to Layer Properties > Labels, select ADMIN2 as the display field, right-click the layer, and click Label Features).
- ghadmin3.shp—Use an Arctic White outline of width 1.5 and a Lime Dust fill color.

Compare the downloaded administrative boundaries with those you already have in the GhanaFlooding geodatabase.

13. **Add the Ghana and Regions feature classes from the GhanaFlooding.gdb to your ArcMap session. (Use the geodatabase provided in the Chapter4 data folder, which is comparable to the one you created in Chapter 3.).**

14. **Click OK to acknowledge that the two layers will be reprojected to match the spatial reference already established in the data frame.**

15. **Optimize the symbology for the Ghana and Regions layers so that you can compare their line definition with the downloaded data.**

The boundary definitions of the Ghana and Regions layers are less precise than those of the downloaded data. But should we replace Ghana and Regions with ghadmin1 and ghadmin2?

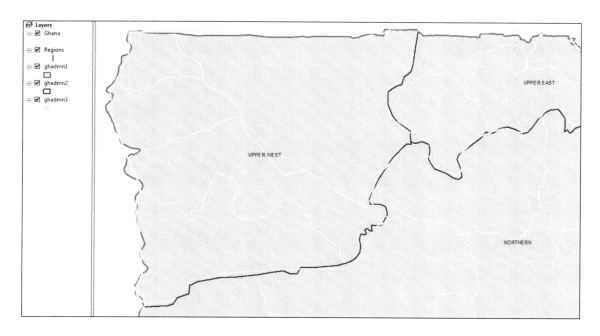

16. Open the attribute tables of the three downloaded shapefiles (Admin Level 1, 2, and 3), and examine their contents.

Do you see why you need to be vigilant about vetting unfamiliar data before adding it your geodatabase? The attribute data of the downloaded files is incorrectly associated with the country of Guinea, a different country that is also in West Africa. Also, the files do not include any population data, which is provided in the Regions layer and obviously very important to humanitarian planning.

It may be worthwhile to merge the Ghana and gadmin1 layers, as well as the Regions and gadmin2 layers, into one definitive first administrative layer boundary and one definitive second administrative layer boundary, respectively. But since we are still relatively unfamiliar with the data, we will not merge the layers now.

Your turn

Using the techniques you learned in chapter 3 and the knowledge about the data you have gained, load all three layers into the Administrative feature dataset of your geodatabase. Rename them AdminLevel1, AdminLevel2, and AdminLevel3. Then use the information provided on the ADDS Web site to ensure that your feature classes have FGDC-compliant metadata.

Exercise 4.2

Reading x,y data into ArcMap

When humanitarian organizations need spatial data that is not available, they generate it themselves. They collect locations of features of interest, such as hazard zones, water points, or evacuation routes, using Global Navigation Satellite System (GNSS) receivers. The United States' GPS is the only fully operational GNSS in use today, although Russia, the European Union, and China are planning to implement several alternative systems over the next decade.

While it is very easy to directly upload data collected from a receiver, oftentimes x,y data is recorded by hand or into nonspatial tables, such as databases or spreadsheets, along with associated notes and attribute data. Supporting a wide range of coordinate systems and units of measurement, ArcGIS enables you to import those x,y coordinates.

Create a data layer from a list of x,y coordinates

From a colleague in the field, you receive a simple database file that contains the x,y (latitudinal and longitudinal) coordinates for several schools destroyed by the flooding. This information must be visualized and integrated into the geodatabase you just created.

1. Start a new map session in ArcMap.

2. Add the DamagedSchools.txt file from the Chapter4 folder's base data.

3. Open the table and view its contents.

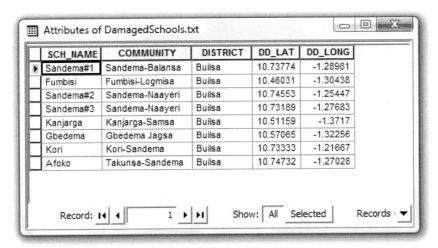

X,y coordinates must be stored in separate columns: the DamagedSchools text file has a latitude column named DD_LAT, and a longitude column named DD_LONG. (DD refers to decimal degree coordinates.)

4. Close the table.

5. On the main menu, go to Tools > Add XY Data.

The Add XY Data tool requires you to specify the source, fields, and coordinate system of your x,y data table.

6. **Populate the tool's dialog box as follows:**

    - Table: DamagedSchools.txt (the default selection)
    - X Field: DD_LONG
    - Y Field: DD_LAT
    - Coordinate system: WGS 1984 (**Hint:** Click the Edit button and open Geographic Coordinate Systems > World folder)

7. **Click OK once your settings appear as follows:**

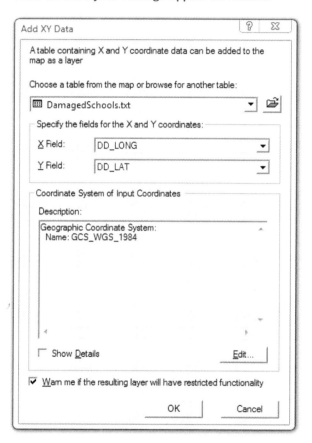

8. **Click OK. If you receive the following message, click OK again. (You will create Object IDs for each of the data records in the next steps.)**

Your DamagedSchools data is now added to the table of contents as a point data layer and projected onto the map display. To give the point layer better context, add the Admin_Level 3 layer and label its features.

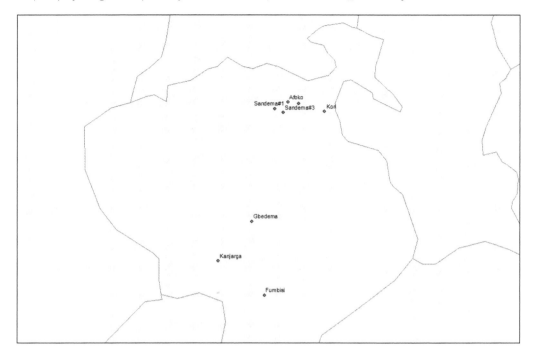

The damaged schools data pertains to one district only, in the northeastern corner of the country. Notice that the new layer is identified in the table of contents as an event. This means that it is a temporary file within the current ArcMap session and not a permanent shapefile or feature class.

9. **Right-click the DamagedSchools.txt Events layer, and go to Data > Export Data.**

10. **Save the new layer as a feature class in the DamageAssessment feature dataset of your geodatabase. You will need to change the default file type from "Shapefile" to "File and Personal Geodatabase feature classes" in order to navigate into your GhanaFlooding geodatabase.**

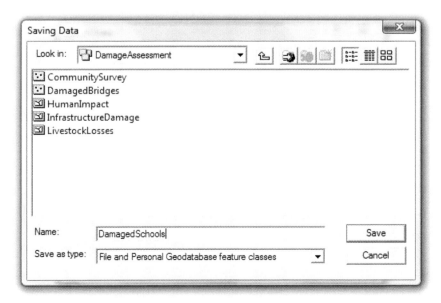

11. **Click Yes when asked if you want to add the exported data to the map as a layer, and confirm that each record of the DamagedSchools feature class has been assigned an object ID.**

Convert x,y data from DMS to DD format

X,y data is not always delivered in a format that is immediately compatible with ArcMap. This is especially true of locations manually extracted from map sheets.

The most traditional way of recording x,y data is in latitude and longitude (lat/long). Lat/long coordinates are often recorded in degrees-minutes-seconds (DMS) format (1 degree = 60 minutes, and 1 minute = 60 seconds). For example, the lat/long coordinates of Accra, Ghana's capital, are 5° 33' 0" N, 0° 12' 0" W.

However, in order for ArcMap to create a point layer from x,y coordinates, it is necessary to convert DMS values into DD (decimal degrees) format. When converted to decimal degrees, latitude values from the southern hemisphere and longitude values from the western hemisphere become negative. For example, Accra's lat/long coordinates become 5.55°, -0.2°.

This part of the exercise will teach you how to convert degrees-minutes-seconds into decimal degrees using Field Calculator.

1. **Add the AffectedCommunities.dbf table from your Basedata folder to your table of contents.**

2. **Open the table.**

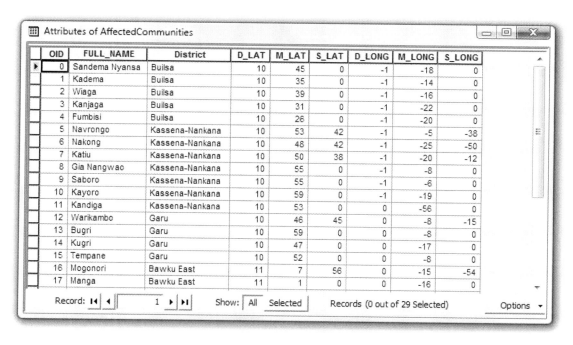

OID	FULL_NAME	District	D_LAT	M_LAT	S_LAT	D_LONG	M_LONG	S_LONG
0	Sandema Nyansa	Builsa	10	45	0	-1	-18	0
1	Kadema	Builsa	10	35	0	-1	-14	0
2	Wiaga	Builsa	10	39	0	-1	-16	0
3	Kanjaga	Builsa	10	31	0	-1	-22	0
4	Fumbisi	Builsa	10	26	0	-1	-20	0
5	Navrongo	Kassena-Nankana	10	53	42	-1	-5	-38
6	Nakong	Kassena-Nankana	10	48	42	-1	-25	-50
7	Katiu	Kassena-Nankana	10	50	38	-1	-20	-12
8	Gia Nangwao	Kassena-Nankana	10	55	0	-1	-8	0
9	Saboro	Kassena-Nankana	10	55	0	-1	-6	0
10	Kayoro	Kassena-Nankana	10	59	0	-1	-19	0
11	Kandiga	Kassena-Nankana	10	53	0	0	-56	0
12	Warikambo	Garu	10	46	45	0	-8	-15
13	Bugri	Garu	10	59	0	0	-8	0
14	Kugri	Garu	10	47	0	0	-17	0
15	Tempane	Garu	10	52	0	0	-8	0
16	Mogonori	Bawku East	11	7	56	0	-15	-54
17	Manga	Bawku East	11	1	0	0	-16	0

Record: |◀ ◀ 1 ▶ ▶| Show: All Selected Records (0 out of 29 Selected) Options ▾

The latitude and longitude information is divided into six separate fields, D_LAT, M_LAT, S_LAT, D_LONG, M_LONG, and S_LONG. In order for the data to be intelligible to ArcMap, you need to combine the first three fields into a single field for latitude, and the last three fields into a single field for longitude. You will then convert those single fields from DMS to DD format.

3. **Click the Options button at the bottom of the table, and then click Add Field.**

4. Type **DMS_LAT** in the Name field.

5. Click the Type drop-down arrow, and then click Text from the list; choose a length of 10.

6. Click OK when your settings appear as follows.

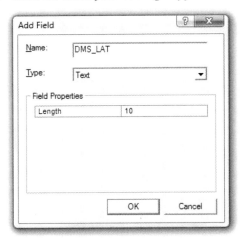

7. Right-click the DMS_LAT field header and select Field Calculator. A Field Calculator warning will advise you that you are about calculate outside of an editing session. Click OK to proceed.

8. Select the Type = String (not the default type Numeric) and, in the dialog box, build the following expression:
   [D_LAT] &" " & [M_LAT] &" " & [S_LAT]

The formula combines the three input strings with a space placed between the quotes "_" to create a single field DMS_LAT that will be formatted like this: DD MM SS.

9. When your Field Calculator appears as follows, click OK.

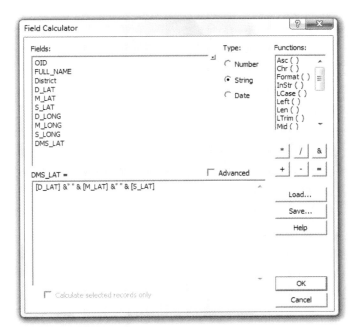

Your DMS_LAT field should now be populated with the combined degree-minute-seconds information.

FULL_NAME	District	D_LAT	M_LAT	S_LAT	D_LONG	M_LONG	S_LONG	DMS_LAT
Sandema Nyansa	Builsa	10	45	0	-1	-18	0	10 45 0
Kadema	Builsa	10	35	0	-1	-14	0	10 35 0
Wiaga	Builsa	10	39	0	-1	-16	0	10 39 0
Kanjaga	Builsa	10	31	0	-1	-22	0	10 31 0
Fumbisi	Builsa	10	26	0	-1	-20	0	10 26 0
Navrongo	Kassena-Nankana	10	53	42	-1	-5	-38	10 53 42
Nakong	Kassena-Nankana	10	48	42	-1	-25	-50	10 48 42
Katiu	Kassena-Nankana	10	50	38	-1	-20	-12	10 50 38
Gia Nangwao	Kassena-Nankana	10	55	0	-1	-8	0	10 55 0
Saboro	Kassena-Nankana	10	55	0	-1	-6	0	10 55 0
Kayoro	Kassena-Nankana	10	59	0	-1	-19	0	10 59 0
Kandiga	Kassena-Nankana	10	53	0	0	-56	0	10 53 0
Warikambo	Garu	10	46	45	0	-8	-15	10 46 45
Bugri	Garu	10	59	0	0	-8	0	10 59 0
Kugri	Garu	10	47	0	0	-17	0	10 47 0
Tempane	Garu	10	52	0	0	-8	0	10 52 0
Mogonori	Bawku East	11	7	56	0	-15	-54	11 7 56

Record: 0 Show: All Selected Records (0 out of 29 Selected) Options

You will now convert the DMS_LAT field into DD format.

10. Add another field to the table.

11. Name the new field DD_LAT, and then make it type Double.

12. Accept the default values for Precision (number of digits that can be stored in the field) and Scale (number of decimal places). The zeros will provide maximum storage of 19 and 11 for a Double field.

When you create new fields in any database, you must specify their type. You selected a 10 character text type for the DMS_LAT field, since it is a nonnumeric string. But the DD_LAT field needs to be numeric. You can store numbers in one of four numeric data types:

- Short integers
- Long integers
- Single-precision floating point numbers, often referred to as "floats"
- Double-precision floating point numbers, commonly called "doubles"

In choosing the data type, first consider the need for whole numbers versus fractional numbers. If you need to store whole numbers, such as 12 or 12,345,678, specify a short or long integer. If you need to store fractional numbers that have decimal places, such as 0.23 or 1234.5678, specify a float or a double.

Secondly, when choosing between a short or long integer, or between a float or double, choose the data type that takes up the least storage space. This will not only minimize the amount of storage required but also improve performance. If you need to store integers between -32,768 and 32,767 only, specify the short integer data type because it takes up only 2 bytes (the long integer data type takes up 4). If you are storing fractional numbers between -3.4E-38 and 1.2E38 only, specify the float data type because it takes up 4 bytes (the double data type takes up 8).

So why did you make the DD_LAT field type Double? In file geodatabases, floats are restricted to 6 digits, whereas doubles have up to 15 digits of precision. When converting DMS to DD format, you need up to 3 digits for the degrees value, another space for a minus (-) sign in case the longitude values are west of Greenwich, and enough decimal place digits to avoid excessive truncation of converted minutes and seconds values.

13. **Assign the DD_LAT a precision of 10 digits (total field length) and a scale of 7 digits (decimal places). Click OK when your settings appear as follows.**

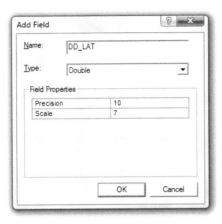

14. **Right-click the DMS_LAT field header, and then select Field Calculator.**

15. **Click Yes if presented with the edit warning, and clear the previously created calculation if it appears in the expression box.**

16. **Check the Advanced check box.**

Note that in the next step you will be asked to copy a lengthy portion of code. Instead of typing this code manually, take advantage of ArcGIS Desktop Help. In the Help search box, type **decimal degrees**. Go to the resulting "Converting Degrees Minutes Seconds Values to Decimal Degrees." You will be able to copy and paste the following code into your Field Calculator wizard.

17. **Paste or type the following code into the expression box:**

```
Dim Degrees as Double
Dim Minutes as Double
Dim Seconds as Double
Dim DMS as Variant
Dim DD as Double

DMS = Split([DMS _ LAT])
Degrees = CDbl(DMS(0))
Minutes = CDbl(DMS(1))Seconds = CDbl(DMS(2))
DD = (Seconds/3600) + (Minutes/60) + Degrees
```

18. **Type the following code into the "DD_LAT =" box at the bottom of the dialog box. (Note that the fourth character in CDbl is a lowercase "L", not the number "1".)**

```
CDbl(DD)
```

19. When the settings appear as follows, click OK.

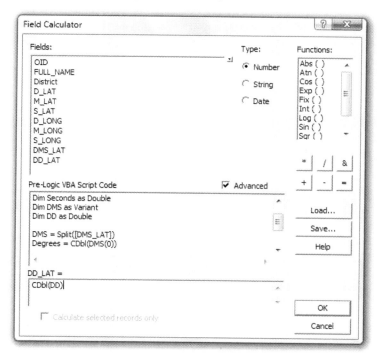

You have now successfully converted your segregated DMS latitude values into a single DD latitude value.

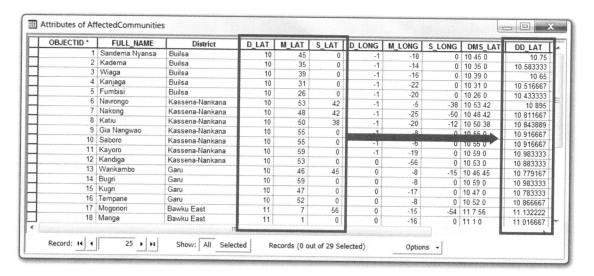

Your turn

Using the method above, convert the D_LONG, M_LONG, and S_LONG field values into a single DD_LONG field value for each record of the Attributes of AffectedCommunities table. **Hint:** You will need to replace DMS_LAT with **DMS_LONG** in the Field Calculator expression code.

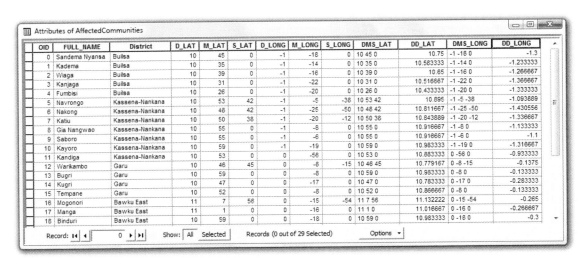

OID	FULL_NAME	District	D_LAT	M_LAT	S_LAT	D_LONG	M_LONG	S_LONG	DMS_LAT	DD_LAT	DMS_LONG	DD_LONG
0	Sandema Nyansa	Builsa	10	45	0	-1	-18	0	10 45 0	10.75	-1 -18 0	-1.3
1	Kadema	Builsa	10	35	0	-1	-14	0	10 35 0	10.583333	-1 -14 0	-1.233333
2	Wiaga	Builsa	10	39	0	-1	-16	0	10 39 0	10.65	-1 -16 0	-1.266667
3	Kanjaga	Builsa	10	31	0	-1	-22	0	10 31 0	10.516667	-1 -22 0	-1.366667
4	Fumbisi	Builsa	10	26	0	-1	-20	0	10 26 0	10.433333	-1 -20 0	-1.333333
5	Navrongo	Kassena-Nankana	10	53	42	-1	-5	-38	10 53 42	10.895	-1 -5 -38	-1.093889
6	Nakong	Kassena-Nankana	10	48	42	-1	-25	-50	10 48 42	10.811667	-1 -25 -50	-1.430556
7	Katiu	Kassena-Nankana	10	50	38	-1	-20	-12	10 50 38	10.843889	-1 -20 -12	-1.336667
8	Gia Nangwao	Kassena-Nankana	10	55	0	-1	-8	0	10 55 0	10.916667	-1 -8 0	-1.133333
9	Saboro	Kassena-Nankana	10	55	0	-1	-6	0	10 55 0	10.916667	-1 -6 0	-1.1
10	Kayoro	Kassena-Nankana	10	59	0	-1	-19	0	10 59 0	10.983333	-1 -19 0	-1.316667
11	Kandiga	Kassena-Nankana	10	53	0	0	-56	0	10 53 0	10.883333	0 -56 0	-0.933333
12	Warikambo	Garu	10	46	45	0	-8	-15	10 46 45	10.779167	0 -8 -15	-0.1375
13	Bugri	Garu	10	59	0	0	-8	0	10 59 0	10.983333	0 -8 0	-0.133333
14	Kugri	Garu	10	47	0	0	-17	0	10 47 0	10.783333	0 -17 0	-0.283333
15	Tempane	Garu	10	52	0	0	-8	0	10 52 0	10.866667	0 -8 0	-0.133333
16	Mogonori	Bawku East	11	7	56	0	-15	-54	11 7 56	11.132222	0 -15 -54	-0.265
17	Manga	Bawku East	11	1	0	0	-16	0	11 1 0	11.016667	0 -16 0	-0.266667
18	Binduri	Bawku East	10	59	0	0	-18	0	10 59 0	10.983333	0 -18 0	-0.3

20. When you have successfully calculated the DD_LAT and DD_LONG field, use the Add XY Data tool to plot your new points in ArcMap.

21. Convert the new Event layer into a feature class within your geodatabase's DamageAssessment feature dataset.

22. Use ArcCatalog to confirm that both of your new layers are listed with the other layers.

23. Complete the required fields of metadata for the two new layers. Because we know very little about the data, indicate that the source, quality, and dates of observation are unknown.

Exercise 4.3

Digitizing features using satellite imagery

The declassification of the technology used to capture fine-resolution imagery from space has been an especially exciting development for humanitarian information managers. Through a variety of commercial and noncommercial mechanisms, data that was restricted to just a handful of military users before the late 1990s is now available for almost any location on earth. This makes it possible to obtain up-to-date information over areas that might otherwise remain very poorly understood. In the context of a humanitarian emergency, satellite imagery has already proven its utility for damage assessment, population monitoring, and many other applications.

In this exercise, you will convert medium-resolution raster data into vector files, delineating the extent of flood inundation in the affected area. You will then use ArcCatalog to build the file structure for a new feature class within your geodatabase.

Create a new (empty) feature class

1. In ArcCatalog, expand the files in your GhanaFlooding geodatabase.

2. Right-click the DamageAssessment feature dataset, and go to New > Feature Class.

3. Name the feature class **FloodExtent**, leave the alias field blank, and choose Polygon Features from the Type drop-down menu. Leave both options for Geometric Properties empty, and then click Next.

4. Accept the default database storage configuration, and click Next.

5. Type **FLOOD_DATE** into an empty row in the Field Name column. Choose Date as the data type.

6. When the settings appear as follows, click Finish.

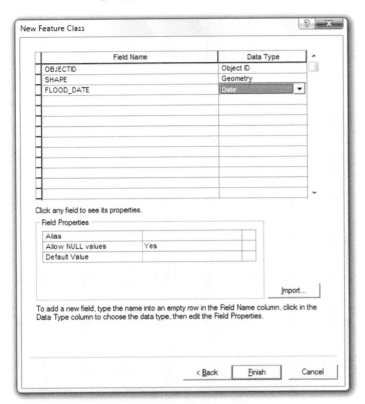

You are now ready to populate the FloodExtent feature class by manually digitizing polygons that delineate the flooded areas visible in your geodatabase's MODIS satellite imagery.

7. **Close ArcCatalog and open the project document GISHUM_C4E3.mxd in ArcMap.**

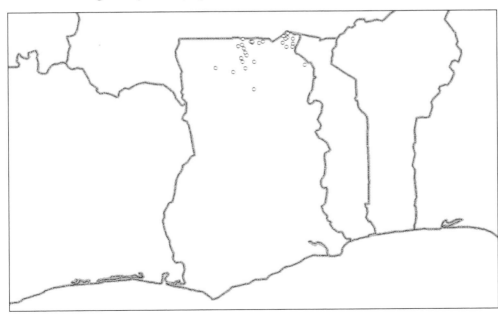

The AffectedCommunities and WestAfricaCountryOutlines layers should appear. If your map display does not look like the preceding figure and the layers appear gray in the table of contents, click the missing layers and reorient them to the actual location of those feature classes in the GhanaFlooding geodatabase (which you originally created in your Chapter4 folder).

Compare spatial data using the Swipe Layer tool

The Swipe Layer tool is used to interactively reveal layers beneath the layer being swiped. This tool makes it easy to see what is underneath a particular layer without having to turn it off in the table of contents.

1. Add the MODIS_PreFlood and MODIS_PostFlood raster datasets to your map display.

2. On the main menu, select Bookmarks > Flooded Region.

3. On the main menu, select View > Toolbars > Effects. The Effects toolbar will be added to your map display.

4. Change the Layer in the toolbar to MODIS_PreFlood.

5. Select the Swipe Layer tool ▼ on the Effects toolbar and, holding down the left mouse button, slide the mouse vertically or horizontally to compare the preflood and postflood MODIS imagery.

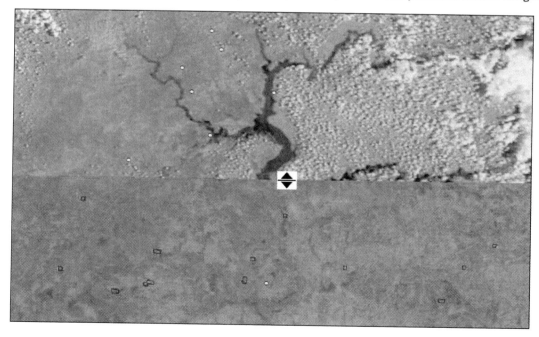

The Swipe Layer tool is very effective for detecting change between any two layers, or two groups of layers, in a map display.

The Effects toolbar offers several other handy features, including brightness, contrast, and transparency adjustment tools. You can reveal hidden layers using the Flicker tool as well. Set the flicker interval you want (in milliseconds) on the Flicker Rate box, and click the 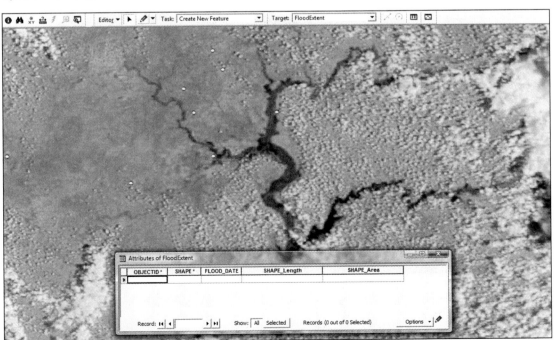 button to begin the flickering of your preflood and postflood images.

(**Reality check:** The small red polygons in the preflood MODIS imagery reveal that it has been edited by a previous user and contains legacy artifacts. While the image is good enough for our purposes, the artifacts would seriously affect quantitative or automated analysis of the MODIS_PreFlood raster dataset. Surprises like this are quite common when working with data that has been used—and altered—by others.)

Digitize vector features from raster data

There are several ways to digitize new features: in this exercise you will employ the "heads up" digitizing method to generate polygon features in your FloodExtent layer based on the MODIS_PostFlood raster dataset.

1. **Add the new FloodExtent feature class to the table of contents.**

2. **Resymbolize the default settings of the FloodExtent to ensure that the features you are about to create are clearly visible against the MODIS imagery.**

3. **Display the Editor toolbar (go to View > Toolbars > Editor, or click the button on the Standard toolbar).**

4. **In the Editor toolbar, click Start Editing; confirm that your task is set to Create New Feature; and set your target layer to FloodExtent.**

5. **Open the (empty) Flood Extent attribute table, and then position it at the bottom of your screen.**

You are now ready to begin digitizing. Rather than taking the time to digitize the entire flood extent, you will digitize just a sample portion.

6. **On the main menu, go to Bookmarks > Digitize 01.**

7. **On the Editor toolbar, click the Sketch** ✎ **tool, and then begin to trace around the edge of the flooded area.**

As you trace around the edge of the shape, be careful to click only once to create vertices. Double-clicking will close the polygon and complete the sketched feature. Although the image is quite pixilated at this scale, try to follow the line of the flooding as closely as possible.

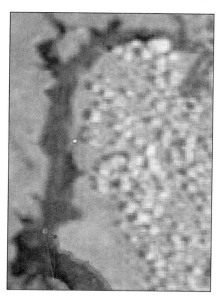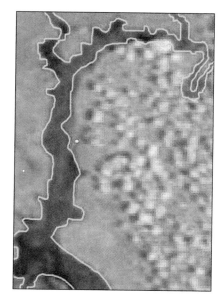

Sketching with a mouse requires patience and practice—and a steady hand! You may wish to zoom and pan frequently to optimize the display. This will not interrupt the sketching process—click the Sketch tool whenever you want to recommence digitizing.

An alternate, more precise method of digitizing complex features is to work at full resolution but in smaller sections of the region. You digitize segments of flood inundation separately, and then merge those separate, contiguous segments into a single flood polygon feature.

8. **Zoom and pan to the region of the image you need to digitize next.**

9. **Starting slightly inside the edge of your last polygon, digitize the next part of the flood inundation feature. (By slightly overlapping contiguous sections, you can be sure that they will merge into one continuous feature later on.)**

10. Continue with this section-by-section method until you have finished digitizing the perimeter of one contiguous, flooded area in the map display.

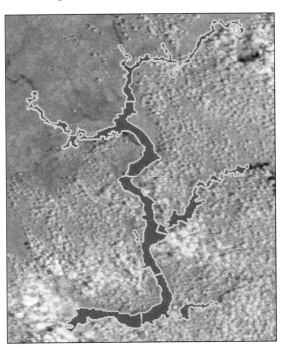

Try to do some or all of the preceding sections, saving your work regularly as you progress.

11. In the FloodExtent attribute table, select all the records you want to merge.

12. On the Editor drop-down menu, click Merge.

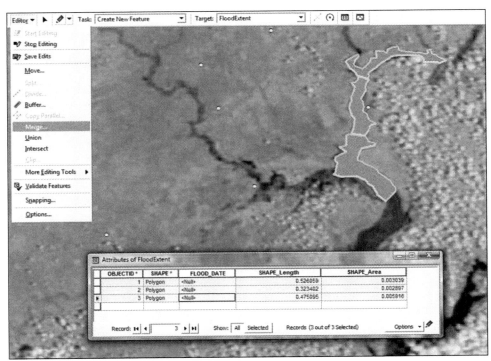

13. Click **OK** to merge all selected records into FloodExtent-1, the first record.

Your digitized sections will be merged on the map and in the attribute table.

14. Type the date of the MODIS_PostFlood imagery into the Date field of the merged record. (You can use dd/mm/yyyy or mm/dd/yyyy formatting to enter the image acquisition date of September 15, 2007.)

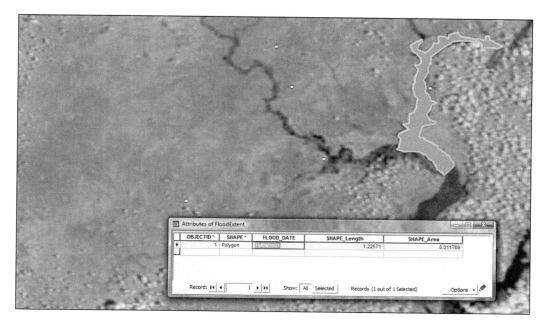

15. Once you are finished creating features for the FloodExtent layer, click Stop Editing on the Editor drop-down menu, and then save your edits.

16. Save the changes to your GISHUM_C4E3.mxd document.

17. Open ArcCatalog, and then create FGDC-compliant metadata for the FloodExtent layer.

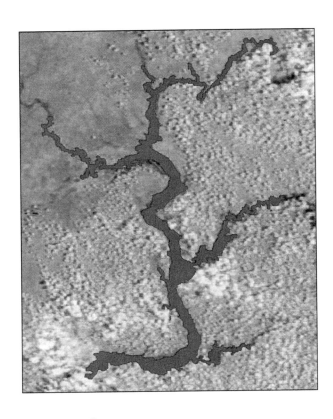

What to turn in

If you are working in a classroom setting with an instructor, submit an electronic copy or a printout of your FloodExtent metadata and FloodExtent layer's geographic coverage.

Exercise 4.4

Sharing data with non-GIS users

You are no doubt aware of the impact of geovisualization tools like ArcGIS Explorer, Google Earth, and Bing Maps (formerly called Microsoft Virtual Earth) since their advent a few years ago. The humanitarian community was especially quick to embrace them to support its basic mapping and reporting needs, and these tools remain a useful complement to full-featured GIS software like ArcGIS. Because many data providers now publish their information in a format different than the native ArcGIS format, it is essential to learn how to export and import information with non-GIS users.

In this exercise, you will learn how to share your data with users of Google Earth and how to use ArcGIS Explorer to present your analysis of the 2007 Ghana floods.

Convert to KML

Keyhole Markup Language (KML) is an XML-based language provided by Google for defining the graphic display of spatial data in applications such as Google Earth and Google Maps. Many humanitarian agencies have embraced KML as their preferred format for exchanging spatial information, and ArcGIS makes it easy for you to export your layers in KML for geovisualization by others.

1.   **Open the GISHUM_C4E3.mxd if you are starting a new ArcMap session.**

2.   **Open ArcToolbox by clicking the ArcToolbox** 🗔 **button in the Standard toolbar.**

3.   **Go to Conversion Tools > To KML.**

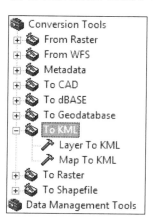

The toolset includes tools that allow you to convert individual data layers or entire maps to KML. We will first convert the FloodExtent feature class (Layer to KML), and then convert the GISHUM_C4E3 map document (Map to KML).

4.   **Double-click the Layer to KML tool.**

5. **Populate the Layer to KML wizard as follows:**

- Layer: FloodExtent
- Output File: Chapter4\GE\GhanaFloodExtent_15092007.kmz
- Layer Output Scale: 1
- Data Content Properties: Return single composite image
- Accept all other default settings.

6. **When your settings appear as follows, click OK.**

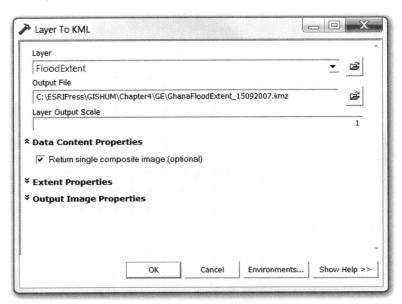

Your Chapter4 data folder now holds a KML file containing a raster of the polygons you digitized in the last exercise. KML files that are zipped (compressed) have a .kmz extension, a native compression utility that can be read by any KML client, including ArcGIS Explorer, ArcGlobe, and Google Earth.

7. **Double-click the Map to KML tool in ArcToolbox.**

8. **Populate the Map to KML wizard as follows:**

- Map Document: Chapter4\GISHUM_C4E3.mxd
- Data Frame: Layers (relevant only if your map display has multiple data frames)
- Output File: Chapter4\GE\GhanaPrePostFloodMap.kmz
- Map Output Scale: 1
- Accept all other default settings.

9. **When your settings appear as follows, click OK.**

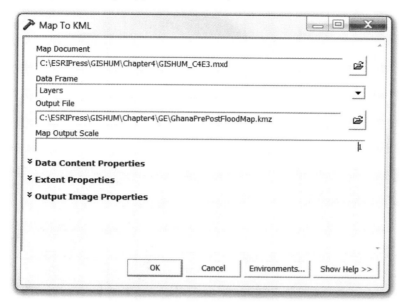

Your exercise 4.3 map document has now been saved in KML as a series of separate layers that can be displayed in any KML-compatible software client. Think of your converted file as a layer package.

Convert from KML

Since many agencies, including the UN, regularly share their data using KML, you may want the ability to import that data into your geodatabase. Visit `http://arcscripts.esri.com/details.asp?dbid=15603` to download an ArcScript designed to allow you to do this.

View data in ArcGIS Explorer

ArcGIS Explorer is a free, downloadable GIS viewer that provides an easy way to explore, visualize, and share spatial information. ArcGIS Explorer adds value to any GIS because it helps you deliver your authoritative data to a broad audience. It is a data-sharing portal for larger-scale GIS, as well as a stand-alone lightweight GIS client.

With ArcGIS Explorer, you can do the following:

- Access ready-to-use ArcGIS Online basemaps and layers
- Fuse local data with map services to create custom maps
- Add photos, reports, videos, and other information to your maps
- Perform spatial analysis (visibility, modeling, proximity search, etc.)

1. Download and install the latest version of ArcGIS Explorer if it is not already installed on your computer. Go to http://www.esri.com/software/arcgis/explorer/index.html for download information, FAQs, and demos.

2. Close ArcMap or ArcCatalog if they are open, and launch ArcGIS Explorer.

3. On the Home ribbon at the top of the window, click Add Contents > Geodatabase Data.

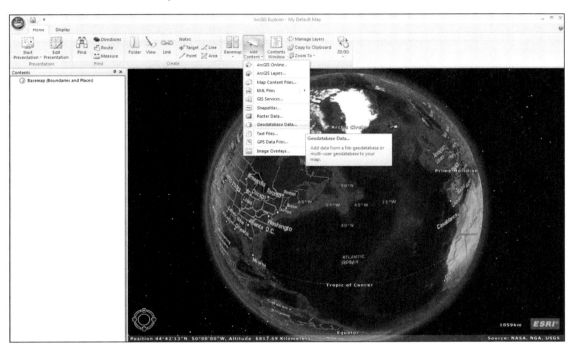

4. Browse to your GhanaFlooding geodatabase, and then click Next.

5. Add one or more of your feature classes and raster datasets to ArcGIS Explorer.

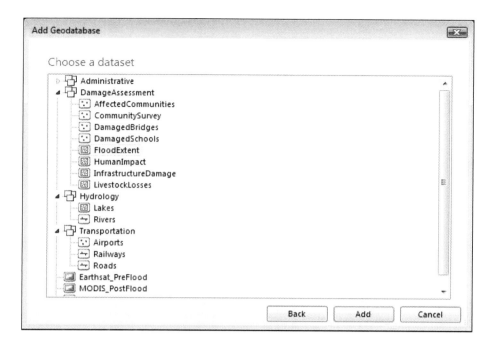

6. **Right-click one of the datasets in the ArcGIS Explorer table of contents, and then click Go To.**

Your display will zoom around the globe toward Ghana, and your selected layers will be visible.

7. **With the same dataset selected, go to the Appearance tab at the top of the window and experiment with the available options for displaying point, lines, and polygon features in ArcGIS Explorer.**

As you can see, quite a few of the basic mapping and querying features of ArcMap are available to users in ArcGIS Explorer as well. ArcGIS Explorer allows you to create maps by incorporating geographic information from a variety of sources, and to add your own notes on top of basemap information. In addition to using ArcGIS Explorer to explore and ask questions about the geography in the map, you can communicate important information to others about that geography with the application by employing your map as a presentation.

There are some options you can set for displaying the presentation. To set the options, click the Options button on the Start Presentation group. This displays the Options dialog box. Try importing your KML files, Microsoft PowerPoint slides, and other content to prepare a short presentation about the 2007 floods in Ghana.

What to turn in

If you are working in a classroom setting with an instructor, deliver a presentation with ArcGIS Explorer using your GhanaFlooding geodatabase (and any other content available to you).

This chapter has given you some essential skills for generating spatial data. In addition to exploiting the vast volumes of georeferenced information available from geodata portals, you have also discovered how to add GPS readings to your geodatabase, extract features from raster imagery, and provide complete metadata. You have learned how to share your ArcGIS layers with the larger humanitarian community via KML. And you have glimpsed the potential in ArcGIS Explorer for serving as an access portal to your geodatabase and an interactive tool for presenting spatial analysis to decision makers.

References

United Nations. 2007. Ghana floods flash appeal 2007. Office for the Coordination of Humanitarian Affairs. New York: United Nations, 26.

Woof, Nigel, Operations Director, MapAction UK. 2008. Personal communication with author between January 17 and March 21.

Chapter 5

Using data models to create your own geodatabase

Contributed by Olivier Cottray

The previous two chapters introduced the file geodatabase and the process required to populate it with data during a humanitarian crisis. While further developing your geodatabase skills, this chapter marks a significant advancement of your ArcGIS skills.

You will create your own geodatabase using the first of a suite of data models being developed under the auspices of the United Nations. You will use an industry-standard framework, the UN Spatial Data Infrastructure for Transport data model (UNSDI-T). Designed to support humanitarian logistics operations, the UNSDI-T provides a common approach to organizing geodatabases that represent features such as roads, ports, and other associated infrastructure.

You will learn to exploit publicly available data models like the UNSDI-T to improve how you manage, store, and exchange your data with other organizations. This lesson prepares you for chapter 7, in which you will use ArcGIS Network Analyst with UNSDI-T to locate warehouses, plan delivery routes, and simulate the effect of route closures in Ethiopia.

What is a GIS data model?

A GIS data model is essentially a framework, built on accepted standards, for modeling and capturing the behavior of real-world objects in a geodatabase (Goodchild 2003). It describes the process of converting tangible ideas or objects into useful and accessible information in a geodatabase.

Since GIS data models provide a template to reduce the barriers to using and sharing geospatial information, they can facilitate more effective GIS implementation and interoperability among organizations. Those that provide humanitarian assistance—United Nations agencies, national governments, donor agencies, research institutes, private sector firms, nongovernmental organizations (NGOs)—are all potential beneficiaries.

The purpose of GIS data models is to provide practical templates for implementing GIS projects for specific applications in various industries. Designed by consortia of experts, these models offer GIS managers a starting point from which to build their internal geodatabases using designs that "really work," allowing them to become productive quickly. By leveraging the schema and thematic groups established by the humanitarian community, you could launch your GIS more efficiently and with greater confidence as well.

GIS data models serve another important purpose by creating a forum for sharing ideas, data, cartographic tools, and expertise among GIS users with common applications of interest. In bringing together a global community of users, the development of data models provides an ideal platform not only to make humanitarian GIS more accessible and effective but also to bring together a global community of users from which to advance the utility of the application more efficiently.

Once a model is completed and available, say as an application schema in Geographic Markup Language (GML), it serves as architecture upon which geodatabase projects can be based. Using the common application model as the basis of their own data model, implementers customize it to suit their purposes. They can also use the model as the basis for dynamically importing and exporting data, dramatically reducing the barriers imposed by nonstandard geodatabase designs.

Existing data models

ESRI began to encourage the development of industry data models around 2000, when it released the ArcGIS 8.x architecture. The shift from cartographic *coverages* to thematic *geodatabases* was, in fact, a transition to object-oriented database management by the GIS industry. The latter now supports a wide range of applications (although data models specifically for the humanitarian and international development community remain elusive). For each of the following data models, ESRI provides case studies, design templates, partner URL links, and online user groups and support (see `http://support.esri.com/datamodels`):

- Address
- Atmospheric
- Basemap
- Biodiversity
- BroadbandStat
- Building Interior Space
- Carbon Footprint
- Census-Administrative Boundaries

- Defense-Intel
- Energy Utilities
- Environmental Regulated Facilities
- Fire Service
- Forestry
- Geology
- GIS for the Nation
- Groundwater
- Health
- Historic Preservation and Archaeology
- Homeland Security
- Hydro
- International Hydrographic Organization
- Land Parcels
- Local Government
- Marine
- National Cadastre
- Petroleum
- Pipeline
- Raster
- Telecommunications
- Transportation
- Water Utilities

Although the investment in the design and maintenance of a data model can be significant, there are tremendous benefits once that data model is adopted throughout an industry. *Designing Geodatabases: Case Studies in GIS Data Modeling* by David Arctur and Michael Zeiler (ESRI Press 2004) is an excellent reference for those interested in understanding and developing GIS data models.

Elements of GIS data model templates

The aforementioned data models are all available as free downloads from the ESRI Support Center. Each includes the following:

- **Thematic layers** composed of the global or regional data pertinent to humanitarian applications in general
- **Spatial representations** consisting of a standard set of feature and image representations with a standard set of names and terms
- **Attributes** to facilitate system interoperability and data sharing as well as consistent use of terminology
- **Integrity rules and spatial relationships**, the foundation of geodatabase design and critical to the success of any GIS application and topological exercise
- **Map layouts** to provide a cartographic standard for communicating information through widely accepted map styles, symbols, and labels
- **Metadata requirements** to support adherence to existing metadata standards, such as ISO and FGDC
- **Extraction guidelines** informing data model users of the rules for data capture, extraction, and sharing
- **Case studies** illustrating the application of the data model in various situations

The need for a humanitarian spatial data infrastructure

To date, the humanitarian community does not have a universal system for defining, naming, or sharing spatial data. This has led to inefficient coordination among the various response agencies: when one organization refers to "aerodromes" (alternatively spelled "airdromes") while others call the same features "airports" or "airfields," how can anyone be sure that these objects are geospatially synonymous? The uncertainty is amplified when one organization's GIS describes "Runway Condition'" with a numeric value ("1"), while another's GIS tracks "Runway Status" with a textual description ("good").

Of course, with time and coordination it is possible to reconcile such semantic differences, but in the context of an unfolding crisis individual agencies often decide to collect their own set of field data even if a near-identical survey was carried out previously by another organization. Since the United Nations is almost always involved during international emergencies, it performs a unique role in the development of GIS data models to support its spatial geodatabase requirements.

Indeed, since its formation in 2000, the United Nations Geographic Information Working Group (UNGIWG) has been developing a UN Spatial Data Infrastructure (UNSDI) comprising data and metadata standards, data-sharing mechanisms, and interoperable geographic data repositories. One member of UNGIWG, the United Nations Joint Logistics Centre (UNJLC), has designed a data model for transport-related datasets as part of this UNSDI initiative. This data model is called the UN Spatial Data Infrastructure for Transport (UNSDI-T).

The UNSDI-T data model is composed of a geodatabase schema (defining layers such as roads, aerodromes, ports, warehouses, railroads, and other logistical infrastructure assets); a set of corresponding data collection forms (ensuring that field data is collected using compatible indicators and values); and a template ESRI File Geodatabase (for users wishing to quickly produce a standardized transportation database compliant with UNSDI specifications).

Read more about UNGIWG and the UNSDI at http://www.ungiwg.org. More information about the UNSDI-T is available from the UNJLC Web site: http://www.logcluster.org/tools/mapcentre/unsdi/unsdi-t-v2.0. (If this link has expired, search the Internet for "UNSDI-T" and find the latest home page for the UNSDI for Transport data model.)

Scenario: A model for food distribution in Ethiopia

In this chapter, you will use the UNSDI-T to learn how to exploit and edit an existing data model in your ArcGIS File Geodatabase. In exercise 5.1, you will use ArcCatalog to import and explore the UNSDI-T geodatabase XML schema using Schema Wizard. Then in exercise 5.2, you will edit your geodatabase structure by setting an x,y coordinate system for a feature dataset and creating a new value domain and new data field. Finally, in exercise 5.3, you will populate your customized geodatabase and create a new feature class using overlay analysis. (Later, in chapter 7, you will use the UNSDI-T data model to optimize food distribution in Ethiopia.)

Layer or attribute	Description
Roads_Export.shp	**Ethiopia road lines (fictionalized)**
Shape_Leng	Road segment
LengthSlope.shp	**Ethiopia slope polygons**
Slope	Slope value, where: 1=Low (less than 15°) 2=Medium (between 15° and 30°) 3=High (more than 30°)

Table 5.1 Data dictionary

Exercise 5.1

Importing geodatabases using ArcCatalog

A geodatabase is a collection of geographic datasets of various types used in ArcGIS and managed in either a common file system folder (termed "file geodatabase"), a Microsoft Access database (a "personal geodatabase"), or a multiuser relational database (ArcGIS Server). The geodatabase is the native data source for ArcGIS and is used for editing and data automation in ArcGIS.

The four main types of data held in geodatabases are feature classes (points, lines, and polygons), descriptive attribute tables, raster datasets, and raster catalogs. To allow for powerful geographic modeling capabilities, the geodatabase also supports a number of other data types and properties. These include relationship classes, topology, subtypes, and network datasets (see chapter 7 for a review of network analysis for an example).

In this exercise, you will learn how to create a file geodatabase from a schema. A schema is a simple file, usually in XML format, that stores information about the definitions, integrity rules, and behaviors for each database element—essentially, how the database is to be structured. You will see how you can easily import a schema into ArcCatalog to automatically generate a predefined file geodatabase. You will then explore some of the main features of the geodatabase in order to better understand its functionality.

Using an Internet browser, navigate to `http://www.logcluster.org/tools/mapcentre/unsdi/unsdi-t-v2.0`, where you will find all the downloadable outputs of the UNSDI-T project. (You can also find out more about history and strategic direction of the UNSDI there.)

Download a geodatabase XML schema

1. **Download the most recent UNSDI-T Comprehensive:Structure file from** `http://www.logcluster.org/tools/mapcentre/unsdi/unsdi-t-v2.0.`

(At the time of publication, UNSDI-T V2.0 was the most current version available.)

2. **Extract the zipped files into the UNSDI-T folder of your Chapter5 data folder.**

Now you will use ArcCatalog to import this schema into a new file geodatabase.

Load and explore a geodatabase model using Schema Wizard

Before you can view the UNSDI-T schema, you will need to add the Schema Wizard tool to your toolbar.

1. **Open ArcCatalog, and then navigate to the folder where you extracted the XML schema.**

2. **On the main menu, go to Tools > Customize.**

3. **Click the Commands tab, and in the Categories frame, select Case Tools. ("Case" stands for computer-aided software engineering, and refers to a family of tools that facilitate software and database development. The Schema Wizard is one such tool, which will automatically convert a schema, or model, into a functional database.)**

4. In the Commands frame, select Schema Wizard, and then drag the icon to your ArcCatalog toolbar.

5. When your dialog box settings appear as follows, click Close.

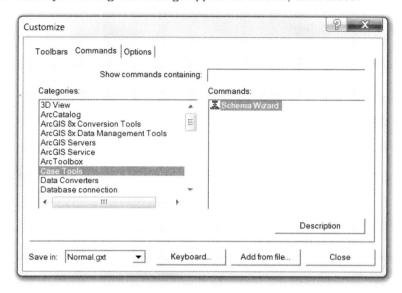

Notice that the tool has been added to your toolbar but is grayed out. The Schema Wizard works by applying a schema to an existing file geodatabase, so you will need to create and select a new file geodatabase before using the tool.

6. Click the UNSDI-T folder to ensure that all of the file contents are listed in the Catalog display.

7. Right-click the UNSDI-T folder, and then go to New > File Geodatabase.

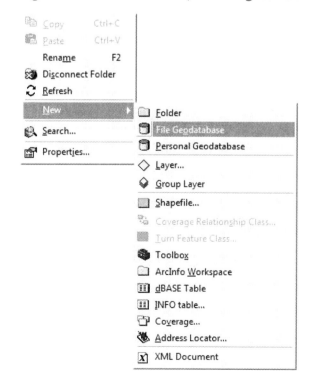

8. Name the new file geodatabase **UNSDI-T**.

9. With the new geodatabase selected, click the Schema Wizard tool ⟟ .

10. Read the introduction screen, then click Next.

11. Click the "Model stored in XML file" radio button, then click Browse.

12. Navigate to UNSDI-T Comprehensive Structure's XML file in your folder, and then click Open.

13. Click Next. Wait for the schema to load.

14. Select "Use values from previous run," and click Next again.

15. Ensure that Transportation dataset is checked, then click Next.

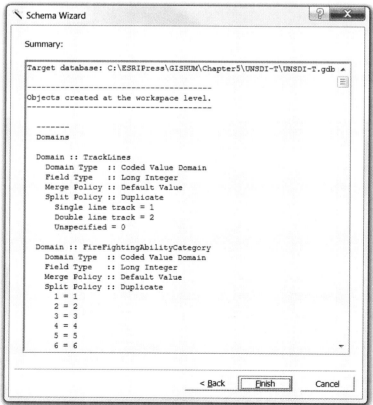

16. **Review the summary of objects to be created, then click Finish to complete the process (this will take a few moments). You may want to view and print the log file for reference before returning to ArcCatalog.**

17. **Right-click UNSDI-T.gdb, and then click Properties.**

18. **Click the Domains tab.**

This window allows the user to set or change the properties for all the value domains held in the database. A domain stores all the possible values that a particular field can take once this domain has been assigned to it.

19. **Scroll down to and select the domain called OperationalStatus.**

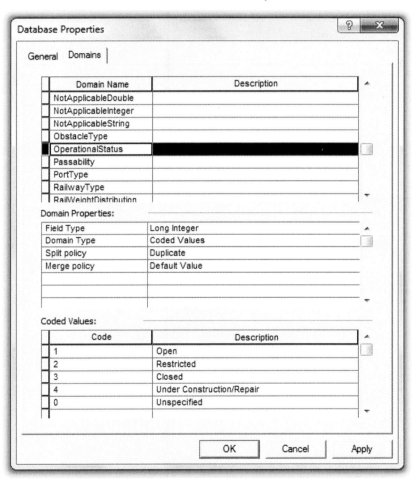

Any field in any table of the geodatabase to which that domain is assigned will accept only the values you can see listed in the bottom frame of the window (Open, Restricted, Closed, Under Construction/Repair, or Unspecified). Any other value entered will return an error and prompt the user to try again. Operational status is a very common and dynamic indicator for logistical assets; having predefined values for it helps speed up the data updating process and prevents discrepancies in terminology and typing errors. Since you own your copy of the geodatabase, you can change the values listed here anytime; this is not recommended, however, if you want your work to remain compatible with other UNSDI-T-compliant projects.

Domain properties

Note the entries listed under "Domain properties" (middle frame). This is where you can set or change the data properties of the values allowed by the domain:

- **Field Type** lets you specify what data type (numeric, text, date) the values listed in the domain must be in order for them to be accepted in the fields to which the domain is assigned. A domain allowing only long integer values cannot be assigned to a text field, for example. You may be wondering why, in that case, the field type for the OperationalStatus domain is set to Long Integer, yet the values we mentioned earlier (Open, Restricted, and so forth) are text values. In fact, this domain saves "coded" values in order to minimize file size. This allows the fields (to which this domain is assigned) to store numeric values in their tables while presenting the user with more intuitive, textual interpretations. Unspecified is stored as the numeric code 0, and Open is stored as the numeric code 1, for example (you can see this value mapping in the two columns of the "Coded values" frame). When interacting with the geodatabase through ArcMap, a user never needs to refer to or even be aware of these numeric codes. As you shall see in later exercises, values are converted automatically during both data entry and data retrieval.

- **Domain Type** refers to whether the domain stores a classified list of coded values (such as the Operational Status classification we just discussed) or a continuous range of values (such as lengths or dates within bounding limits).

- **Split policy** determines what value will be assigned to the line or polygon features resulting from splitting an original line or polygon feature. In the case of coded values, the geodatabase can either assign duplicates of the original feature's value to the resulting features, or the default value set during the definition of the field itself. In the case of range values, the geodatabase can also be set to calculate a ratio of the original value corresponding to the respective lengths or areas of the resulting features.

- **Merge policy** refers to the effect on attribute values when two lines or polygons are merged. When this happens, the value can only be set to the field's default value in the case of coded value domains. In the case of range domains, the value can be set to the default value, the sum of the two constituent parts, or a weighted average of the two constituent parts.

20. Click OK to close the Database Properties window, and then display the contents of UNSDI-T.gdb (either by double-clicking the geodatabase in the contents tab or by clicking it once in the left Table of Contents frame).

Contents	Preview	Metadata	

Name	Type
Transportation	File Geodatabase Feature ...
Berths	File Geodatabase Table
BerthsEquipments	File Geodatabase Table
CompRel_Adrm_TolArea	File Geodatabase Relation...
CompRel_Bridge_Span	File Geodatabase Relation...
CompRel_Port_PrtEntrance	File Geodatabase Relation...
CompRel_WhsCpd_WhsC...	File Geodatabase Relation...
CompRel_WhsCpd_WhsU...	File Geodatabase Relation...
PrtEntrances	File Geodatabase Table
Rel_Berth_BerthEqpt	File Geodatabase Relation...
Rel_Port_Berth	File Geodatabase Relation...
Rel_Port_BerthEqpt	File Geodatabase Relation...
Rel_Source_Adrm	File Geodatabase Relation...
Rel_Source_Bridge	File Geodatabase Relation...
Rel_Source_EntryPoint	File Geodatabase Relation...
Rel_Source_FerryCrossing	File Geodatabase Relation...
Rel_Source_FuelSupplyPo...	File Geodatabase Relation...
Rel_Source_Obstacle	File Geodatabase Relation...
Rel_Source_Port	File Geodatabase Relation...
Rel_Source_Rd	File Geodatabase Relation...
Rel_Source_Rlw	File Geodatabase Relation...
Rel_Source_Stn	File Geodatabase Relation...
Rel_Source_Tunnel	File Geodatabase Relation...
Rel_Source_WhsCpd	File Geodatabase Relation...
Rel_Source_Wtw	File Geodatabase Relation...
Rel_WhsCpd_Berth	File Geodatabase Relation...
Rel_WhsCpd_TempCtrlunit	File Geodatabase Relation...
Rel_WhsUnit_WhsUnitDoor	File Geodatabase Relation...
Sources	File Geodatabase Table
Spans	File Geodatabase Table
TempCtrlUnits	File Geodatabase Table

A list of geodatabase elements appears in the catalog display, including a feature dataset ⬚ (Transportation), a number of attribute tables ▦ (Berths, BerthsEquipments, etc.), and relationship classes ⬚ (CompRel_Adrm_TolArea, CompRel_Bridge_Span, etc.).

Understand geodatabase definitions

A **dataset** is simply a container grouping features with similar properties. If you right-click Transportation and open its properties, you see that you can inspect or set a geographic projection, for example. All feature classes created in this dataset inherit this projection without the user having to define it explicitly for each of them. The same inheritance principle applies for all properties that you can set on the dataset.

An **attribute table** contains additional attributes used to provide a more detailed description of a feature class. An aerodrome held in a point feature class, for example, might have several takeoff and landing (TOL) areas, each of them with different characteristics (orientation, length, surface, etc.). These TOL details are stored in the TolAreas attribute table and are linked to the Aerodromes feature class through a relationship class.

A **relationship class**, such as CompRel_Adrm_TolArea, is a database element defining the relationship between two other elements. (For a thorough discussion of relational database principles, see ArcGIS Help.)

1. **Double-click the Transportation feature dataset, right-click the Aerodromes feature class, and view its properties.**

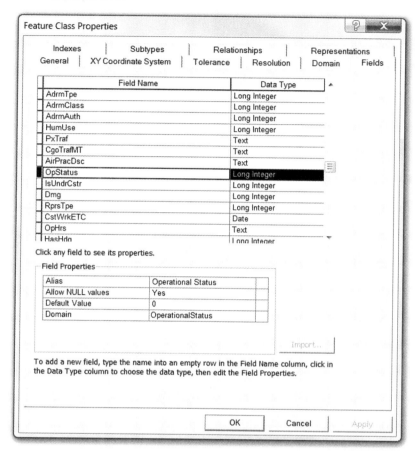

Your turn

Select some of the feature classes in the dataset and explore their properties. What other feature classes have fields to which the OperationalStatus value domain has been assigned?

Having familiarized yourself with some of the main components of a geodatabase, now you will learn how to change some of its properties and adapt its structure to suit your specific information requirements.

Exercise 5.2

Editing geodatabase structure

The UNSDI-T geodatabase structure is intended to provide a comprehensive set of predefined indicators applicable to the majority of humanitarian uses. However, it cannot, and should not, attempt to capture all possible user needs. Individual agencies can delete and add indicators according to their specific information requirements; those UNSDI-T elements that they preserve ensure that their geodatabase is compatible with other UNSDI-T users.

For this exercise, imagine that the nature of the operation requires us to keep track of landslide risk on certain roads. Landslide risk indicators are not included in the standard UNSDI-T package, so an attribute will have to be added to the road feature class. Also, in order to avoid typing errors with potentially dangerous consequences, and to provide a consistent classification throughout the operation, you will create a value domain containing all relevant values of landslide risk.

Set the x,y coordinate system for a feature dataset

1. **In ArcCatalog Catalog tree, expand the UNSDI-T.gdb in your Chapter5 folder.**

2. **Right-click Transportation, and then open its properties.**

3. **Click the XY Coordinate System tab.**

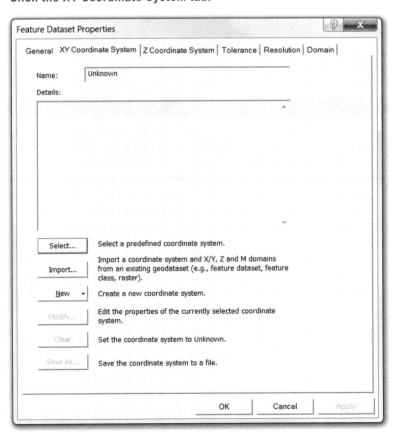

4. Click Select to choose a predefined coordinate system.

5. Go to Geographic Coordinate Systems > World > WGS 1984.prj.

6. When your settings appear as follows, click Add.

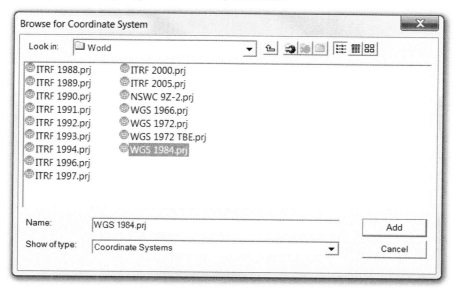

The parameters of the selected projection now appear in the details frame of the XY Coordinate System tab of the dataset properties.

7. Click OK.

The Transportation dataset has now been assigned a coordinate system. Let's examine the feature classes in the dataset.

8. Double-click Transportation.

You are now presented with a list of feature classes and a few more relationship classes. Note that the icons for feature classes are very similar to those used for shapefiles and allow you to immediately recognize their geometry. This feature dataset contains only point and line feature classes.

9. Right-click Aerodromes, and then open its properties.

10. Click the XY Coordinate System tab. Notice that the Aerodromes feature class has been assigned the same geographic projection as its parent dataset.

Notice also that all buttons allowing you to change this projection have been disabled. If you want to define a different projection for Aerodromes, you will need to reproject it into a different dataset with your desired projection or to a stand-alone shapefile.

11. Click the Fields tab.

12. Scroll down to and highlight the field named OpStatus.

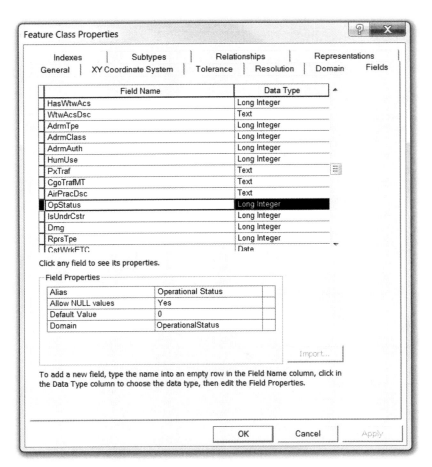

In the Field Properties box, notice that the field's alias is set to Operational Status. Most databases do not allow long field names or names containing spaces and other reserved characters, so truncated names are used instead (such as OpStatus). An alias is set to present the user with a more readable name when interacting with the database.

Note that the field's default value is set to 0 and that its assigned domain is OperationalStatus. You may remember that the value 0 in the OperationalStatus value domain definition represents Unspecified. So any new aerodrome feature added to the feature class will start off with an Unspecified operational status by default.

13. Click OK to close the window.

Create a new value domain

1. **Right-click UNSDI-T.gdb, and then open its properties.**

2. **Click the Domain tab.**

3. **Scroll to the bottom of the Domain Name list, and then select the next empty cell.**

4. **Type Risk in the empty cell.**

5. Under Domain Properties, make the following specifications for the new value domain:

- Field Type: Short Integer
- Domain Type: Coded Values
- Split policy: Duplicate
- Merge policy: Default Value

6. Under Coded Values, make the following specifications about risk levels:

- 0: Unspecified
- 1: Low
- 2: Medium
- 3: High

7. When your settings appear as follows, click OK.

You have just created a domain that will provide data editors with a predefined list of risk values. The values will be stored in the database as numbers (0, 1, 2, and 3) but displayed as more descriptive text (Unspecified, Low, Medium, and High, respectively).

If a road segment is split, the value of the original segment will be allocated to the new segments (Split policy). If two road segments are merged, the new merged segment will be allocated the default value defined for that field (Merge policy—you will set this default value later in this exercise). Giving this value domain a generic name such as "Risk" gives us the flexibility of assigning it to any field in the geodatabase that describes any type of risk. For example, along with a field storing landslide risks for the Roads feature class, we might imagine a field for banditry risk. It would make sense to assign the Risk value domain to this field rather than create an entirely new one.

Important: When publishing a map denoting any type of risk, include a disclaimer or recommendation to users, such as, "Check with local authorities for the most up-to-date risk situation." Due to the dynamic nature of humanitarian operations, maps are almost always obsolete the moment they are published and should only be used for guidance, not as a guarantee of safety.

Create a new data field

Now you will add a new field, called LndSldRsk, to the Roads Feature Class.

1. **From within the Transportation dataset, right-click Roads, and then open its properties.**

2. **Click the Fields tab, then scroll to the bottom of the Field Name list.**

3. **In the empty cell under Shape_Length, type LndSldRsk.**

4. **Set its Data Type to Short Integer.**

5. **In the Field Properties frame, make the following specifications:**

   - Alias: Landslide Risk
   - Allow NULL values: Yes
   - Default Value: 0
   - Domain: Risk

Risk is the only available value, since this is the only domain set to take short integer values.

6. When your settings appear as follows, click OK.

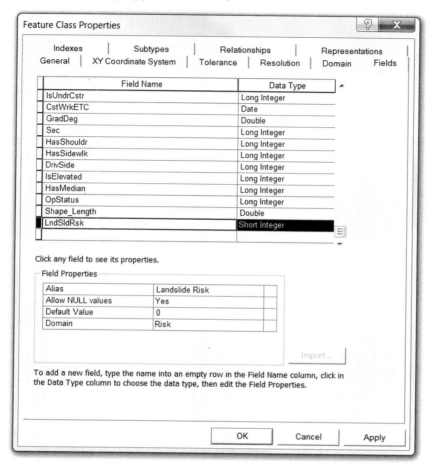

You have just created a new field in the Roads feature class that will accept only values listed in the Risk domain. Moreover, newly created road features will be assigned the default value of 0 (code for Unspecified) until reliable information is acquired, allowing the user to assign a specific landslide risk value.

In the next exercise, you will learn how to load data into the geodatabase using ArcMap. After that comes the opportunity to see for yourself the behavior of the field you just created.

Exercise 5.3

Populating geodatabases

A common data structure simplifies data exchange among organizations and saves time. Once adopted and implemented, a schema compatible with an industry standard (such as the UNSDI-T) enables you to populate your geodatabase easily with data provided by other agencies that also conforms to this standard. Once you have loaded data into your geodatabase, you can focus on data quality analysis and topological cleaning activities, instead of on the lengthy process of interpreting and translating attribute values.

In this exercise, imagine that a partner agency has just sent you a road dataset for Ethiopia. It is a shapefile that was clipped and exported from the agency's large UNSDI-T-compliant geodatabase of the Horn of Africa. You will upload the data into your own lighter geodatabase, then update the new Landslide Risk field.

Load data into a feature dataset

1. If not already opened, open ArcCatalog, and then navigate to UNSDI-T.gdb in your Chapter5 folder.

2. Select the layer Roads_Export.shp.

3. Click the Preview tab, and then select Table in the Preview drop-down menu (bottom of the Preview frame).

If you scroll along the shapefile's attributes, you will notice that they all display numeric values (mostly "0" in this case). This is because shapefiles do not support coded value domains. The link between the numbers and their corresponding textual description (defined in the original geodatabase's Domains property) was therefore lost when the data was exported to shapefile. All you are left with are the original numeric values as stored by the geodatabase. Loading the shapefile's records into your UNSDI-T-compliant geodatabase will reestablish that link; this is what you will do next.

4. Click the Contents tab, and then navigate to the geodatabase UNSDI-T.gdb in your Chapter5 folder.

5. Open the Transportation feature dataset.

6. Right-click the Roads feature class.

7. Select Load > Load Data.

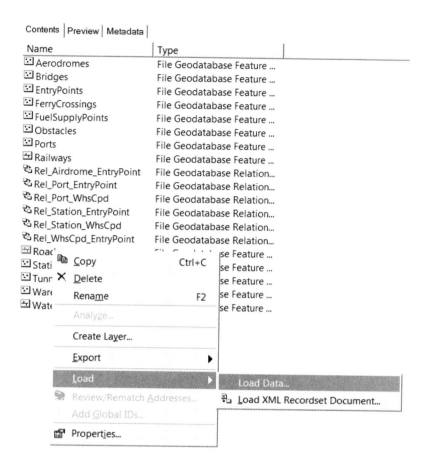

8. Read the introduction screen, then click Next.

9. From the Input data text box, browse to the Roads_Export file.

10. Navigate to C:\ESRIPress\GISHUM\Chapter5, and then select Roads_Export.shp.

11. Click Open.

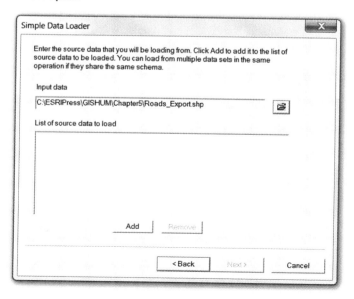

12. Click Add and then Next.

13. Observe the default option "I do not want to add all features to a subtype," then click Next.

14. Click Next to accept the default feature attributes.

15. Check that "Load all of the source data" is selected, then click Next.

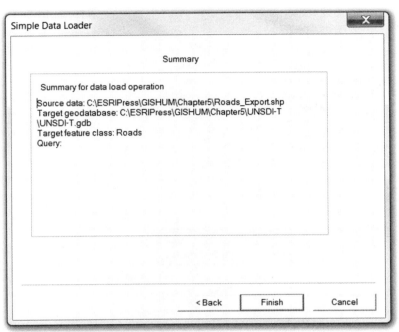

16. Click Finish.

17. Select the Roads feature class in the Catalog tree, and then click the Preview tab.

18. Select Table in the Preview drop-down menu, and then scroll right along the feature class fields.

Notice that most of the fields contain <null> values, since they did not exist in the Roads_Export shapefile and now need to be populated. In the interim, you will create default values that communicate that all values are "Unspecified."

19. Close ArcCatalog, and then open a new empty map in ArcMap.

20. Add the Roads feature class by clicking the Add Data ⬇ button and navigating to the Transportation dataset in your UNSDI-T geodatabase.

21. Open the Attributes of Roads table.

22. Right-click the Existence field header, and then open the Field Calculator.

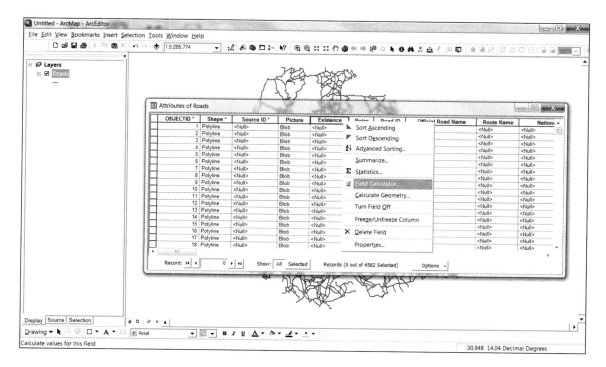

23. Type the number 0 into the Field Calculator's expression box, then click OK.

Although the actual field value in the Roads feature class is recorded as a numeric value, the geodatabase's value domain translates these to text to make it clear that Existence values are unspecified and still need to be added.

Your turn

Change the remaining field values from Null to Unspecified using the preceding process.

24. Close the Roads attribute table.

Use category symbols to illustrate unique attribute values

You have completed the process of uploading data into your geodatabase. You will now populate the Landslide Risk field in ArcMap based on a simple slope model.

1.   Add to your map display the Slope shapefile, located in the Chapter 5 data folder.

This layer was derived from freely available Shuttle Radar Topography Mission (SRTM) elevation data for Ethiopia. A slope raster layer was generated using ArcGIS Spatial Analyst, then classified into low, medium, and high slope categories. The medium and high slope areas were finally converted to the generalized polygons you see here. A value of 3 in the shapefile's Slope field indicates areas where slope is greater than 30 degrees (high); a value of 2 indicates areas of slope between 15 and 30 degrees (medium); a value of 1 indicates areas of slope less than 15 degrees.

These values serve as very simplified indicators of landslide risk for the purposes of this exercise (a reliable assessment of landslide risk requires considering many other factors). In the exercise, roads intersecting areas of medium slope are given a medium landslide risk value; roads intersecting high-slope areas receive a high risk value.

2. **Right-click Slope in the table of contents, and then go to Properties > Symbology > Categories > Unique values.**

3. **Select Slope in the Value Field drop-down menu.**

4. **Click Add All Values.**

5. **Change the symbol colors to turn off outlines and apply the following fill colors:**

    *   Lemongrass Green for Slope = 1
    *   Seville Orange for Slope = 2
    *   Mars Red for Slope = 3

6. **Clear the <all other values> check box at the top of the Symbol field.**

7. **When your settings appear as follows, click OK.**

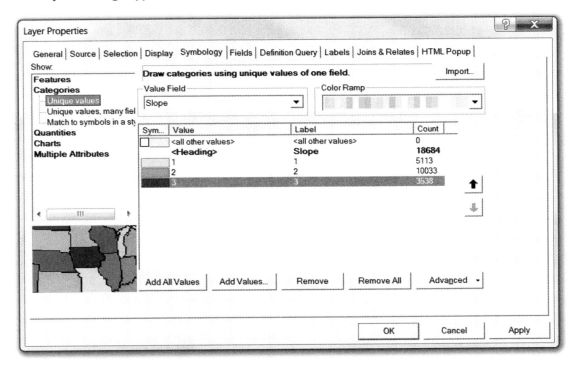

Create a new feature class using overlay analysis

Zoom in on the map. Select a few road segments that intersect slope polygons and you'll find many of them extending into areas where slope is not significant. In order to classify landslide risk along only those portions in medium- or high-slope areas, first you need to split the lines where they intersect the polygons. You will do this using the Intersect tool, part of the Overlay toolset provided in the Spatial Analysis toolbox of ArcToolbox. (Overlays are a form of GIS-based analysis called "transformations," which you will learn more about in chapter 6.)

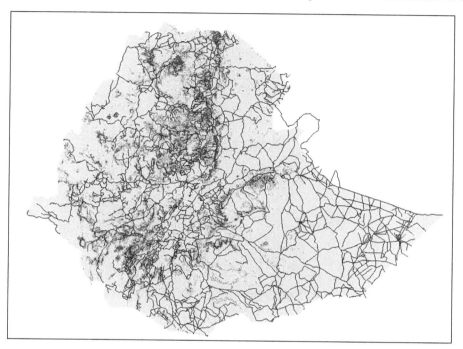

1. **Open ArcToolbox.**

2. **Go to Analysis Tools > Overlay > Intersect.**

3. **Populate the Intersect wizard's dialog box as follows:**

   - Input Features: Roads feature class and Slope shapefile
   - Output Feature Class: Rds_Slope within your Transportation feature dataset
   - JoinAttributes: ALL

4. **When your settings appear as follows, click OK.**

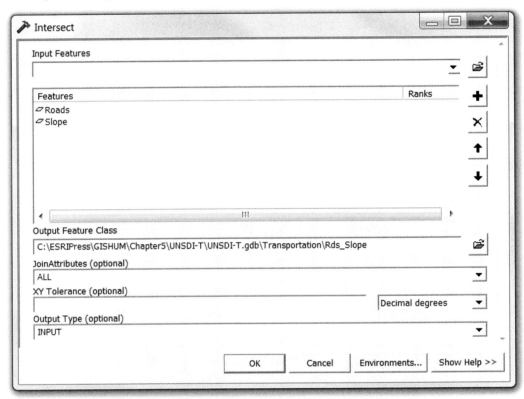

You have now generated a new roads layer split at its intersections with the Slope polygons.

5. **Open the Rds_Slope layer's attribute table, and then scroll to the far right until you see the Slope field.**

The new feature class also contains a new Slope attribute populated according to which polygon category it lies within.

6. **Save your work as a new map document called GISHUM_C5E3.mxd in your Chapter 5 exercise data folder.**

Select feature class records by attribute value

Now select all roads where Slope = 1 and set their Landslide Risk field to Low; then repeat the process, setting Landslide Risk to Medium where Slope = 2, and to High where Slope = 3.

1. **If the Editor toolbar is not visible, on the main toolbar, go to View > Toolbars > Editor, or click the Editor icon ⁚ℓ on the Standard toolbar.**

2. **Click the Editor drop-down menu, and then click Start Editing.**

3. **Under "Which folder or database do you want to edit data from?", highlight the file geodatabase UNSDI-T.gdb. Click OK. This ensures that the Rds_Slope feature class is made available for editing.**

4. Right-click Rds_Slope in the table of contents, and then click Open Attribute Table.

5. Scroll along the fields to the right. Notice that most field headings are now white rather than gray; this indicates that these fields are currently editable. Notice also that the field headings are displaying their aliases rather than their truncated names.

6. Scroll all the way right to find the Landslide Risk field.

7. All records should currently be set to <Null>. However, if you click inside any cell a drop-down list appears containing the values you defined in the Risk value domain. Selecting any of these values populates the record accordingly.

Operational Status	Landslide Risk	FID_Slope	ID	Slope	Shape_Length
<Null>	<Null>	18682	18683	1	0.008039
<Null>	<Null>	18682	18683	1	0.504792
<Null>	<Null>	18682	18683	1	0.543579
<Null>	<Null>	18682	18683	1	0.012984
<Null>	<Null>	18682	18683	1	0.006315
<Null>	<Null>	18505	18506	1	0.184645
<Null>	<Null>	18682	18683	1	0.023996
<Null>	<Null>	18682	18683	1	0.016104
<Null>	<Null>	18682	18683	1	0.036965
<Null>	<Null> ▼	18682	18683	1	0.060618
<Null>	<Null>	18682	18683	1	0.008302
<Null>	Unspecified	18682	18683	1	0.002307
<Null>	Low	18682	18683	1	0.261809
<Null>	Medium	18682	18683	1	0.107965
<Null>	High	18682	18683	1	0.072941
<Null>	<Null>	18505	18506	1	0.037424
<Null>	<Null>	18682	18683	1	0.379418
<Null>	<Null>	18505	18506	1	0.005283

It would be impractical to change so many records manually one by one. Instead, you will select all roads that intersect high-slope areas; change all their values in a single procedure; and repeat for roads intersecting medium- and low-slope areas.

8. From the main menu, click Selection, then Select By Attributes.

9. Under Layer, select Rds_Slope.

10. Scroll down the field list and double-click "Slope".

11. Click the = sign, then type 1 so that the completed query reads: "Slope" = 1.

12. Click OK when your settings appear as follows.

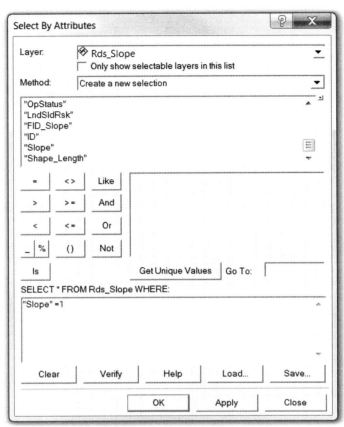

13. Click the Attributes button 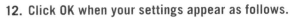 on the Editor toolbar.

14. A window appears with a list of all selected records on the left and the field values of the first highlighted record. You could use this to change the records' values one by one, but you can also set a value for all records of Rds_Slope in one go.

15. Click Rds_Slope at the top of the record list in the left-hand frame to highlight it.

16. Scroll down to the Landslide Risk field, and then click it to highlight it.

17. Point the cursor within the highlighted area, below the Value column, to bring up the value domain drop-down list.

18. Select Low.

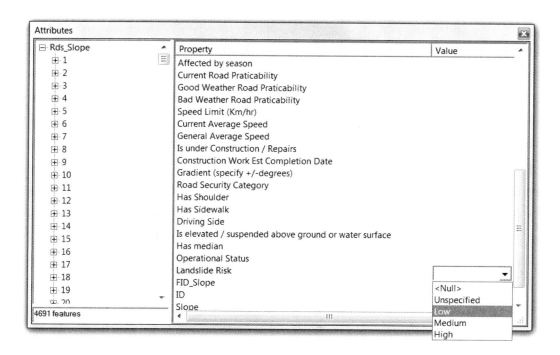

All selected records have now been allocated a Low landslide risk value.

19. Repeat the process for roads crossing areas of medium and high slope, that is, slope values equal to 2 and 3, respectively.

20. Once all values have been calculated for High and Medium landslide risk, click the Editor drop-down button, and then click Stop Editing.

21. Click Yes to save your Edits.

22. Symbolize the Rds_Slope layer according to Landslide Risk values (you may wish to keep <all other values> checked in order to visualize those road segments that still contain NULL values). Use the same color scheme as you did to symbolize slope values in the last exercise, and click the Count field header to ensure that all road segments were correctly classified according to Landslide Risk.

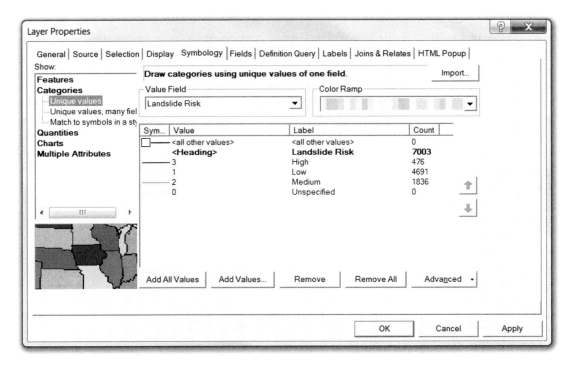

23. Clear the Slope layer check box to better visualize the result of your work.

24. The map should now look as follows.

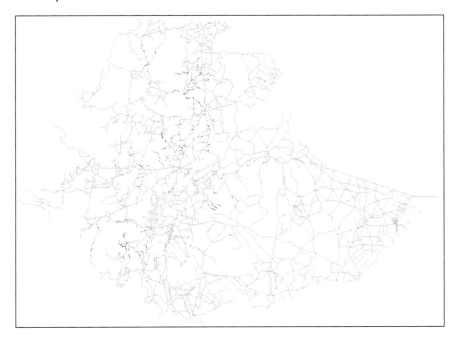

This chapter finishes your foundational training in geodatabases and marks the completion of the first half of this tutorial. Now you are ready to explore the full potential of spatial data, and to help you do so, the remainder of this book is devoted to analytical applications of ArcGIS.

Assignment

Visit the ESRI Support Center and download a data model of interest to you or your organization. Begin by exploring the schema in ArcCatalog and associated supporting documentation. Then edit the schema as required and load your own data into the model.

References

Goodchild, Michael F. 2003. Progress report: The health geodatabase model. University of California at Santa Barbara, May 22.

Chapter 6

Managing hazardous operations

The preceding chapters offered you experience using skills in cartography and data management that are essential to any humanitarian GIS specialist. The remainder of this tutorial is devoted to expanding your ability to exploit ArcGIS for analysis and to guide decision makers who depend upon you for "geo-intelligence." You'll discover that the analytical utility of spatial data offers tremendous potential at the field and headquarters level of any organization.

In this chapter, you will learn how GIS can be used to manage hazards to humanitarian operations. One common category of such hazards is Explosive Remnants of War (ERW), which includes land mines, cluster bombs, and unexploded and abandoned ordnance (UXO). Far too often, contaminated regions remain hazardous long after combat has ended. Understanding the location of such dangerous areas in relation to areas of livelihood and critical infrastructure is a crucial step toward reducing ERW-related risk.

This chapter investigates how hazardous operations can be managed using some of the most common types of spatial data analysis tools: overlay analysis, proximity analysis, extraction, and tabular analysis.

Spatial analysis of the aftermath of war

At the end of the Kosovo War in mid-1999, nearly a million refugees returned home to the aftermath of more than a year of massive warfare. Large areas of Kosovo had been contaminated by cluster bombs, land mines, and unexploded ordnance, some of which resulted from a bombing campaign by NATO forces between March and June 1999. In order to facilitate repatriation and reconstruction activities, the United Nations Mission in Kosovo established a Mine Action Coordination Center (MACC) to conceptualize and develop a work plan for managing ERW hazards throughout the region.

This mission presented an unusual challenge, and unprecedented opportunity, for the MACC staff. The challenge was immense pressure from donors, nongovernmental organizations, local stakeholders, and NATO members to return affected populations to their homes quickly but safely. A thorough, field-level risk assessment, the preferred and conventional practice at that time, would have been very difficult to conduct. The opportunity came not on the ground but from having access to NATO bomb campaign records, military and civilian maps, and modern GIS technologies: the MACC staff could perform their mission faster and more efficiently using spatial analysis.

Generating spatial data for ERW hazard management

The MACC began by designating the locations of minefields, unexploded ordnance, and cluster bomb strikes as danger areas, then collecting and importing various spatial datasets for all records of these. GIS technicians had to scavenge from many organizations and sort through thousands of records of varying reliability in order to generate this spatial data.

As mentioned earlier, collecting reliable data—with the assurance that it's suitable and complete—is a major challenge during a humanitarian crisis. Generating data takes time, perseverance, and "data diplomacy." Then, after your persuasive and persistent efforts pay off, an even more formidable obstacle awaits you: trying to use multiple datasets from multiple sources. Building a solid geodatabase from highly heterogeneous data is no small task! (Fortunately, in this chapter that work has already been done for you.)

The MACC's method of analysis included these three steps:

1. Define the extent and severity of the hazard.
2. Identify safe areas.
3. Set priorities for hazard mitigation.

Step 1: Define the extent and severity of the hazard. After examining the spatial relationships between available data, the MACC staff realized that many data records were redundant. That would not have been as obvious had they been comparing the same data nonspatially. By overlaying various data layers, the staff was able to isolate unique records, improve existing records, and create a better assessment of the extent and severity of the ERW hazard.

Step 2: Identify safe areas. After careful consideration of the civilian and humanitarian requirements for movement within Kosovo, the MACC designated all areas within 500 meters of a town and 200 meters of roads to be treated as safety zones, with extra emphasis placed upon areas used to cultivate or find food (farmland and woodland). This prioritized the management of ERW hazards in and near highly trafficked areas around the country.

Step 3: Set priorities for hazard mitigation. The MACC staff then ranked each district in Kosovo by its level of ERW contamination and by the percentage of each district that needed to be deemed "safe" according to the criteria previously stated. They chose to use a simple scheme based on median values to prevent extremely low and high values from skewing hazard mitigation priorities.

Based upon this analysis, ERW contamination was ranked according to the median value of contamination for all 327 districts in Kosovo, which was 1.3 percent. Anything less than the median value was defined as Low, and values above the median were ranked Medium or High, as indicated in table 6.1.

Level description	Definition
None	0 percent
Low	<1.3 percent (up to the median value of contamination)
Medium	1.3–6.5 percent (between one and five times the median value of contamination)
High	>6.5 percent (more than five times the median value of contamination)

Table 6.1   Level of contamination

These techniques for hazard management were possible because of the availability of modern GIS. MACC staff members are widely recognized as pioneers in the use of GIS for humanitarian operations planning, and many are still actively involved in the fields of humanitarian assistance and GIS. (See "Additional reading" at the end of this chapter.)

Scenario: Land mine management in postwar Kosovo

In this chapter you will analyze a geodatabase of Kosovo following the same basic methodology used by the MACC in the aftermath of the Kosovo War. In exercise 6.1, you perform a series of spatial transformations to calculate the total area contaminated by ERW hazards, and the relative level of contamination by district. In exercise 6.2, you manipulate the tabular values of ERW contamination to classify districts according to the categories defined in table 6.1. Your spatial transformation skills will get a workout in the final two exercises as you develop guidelines for ERW response teams on how to prioritize clearance operations. By the end of the chapter, you will be ready to produce a thematic map for two of the most contaminated municipalities of Kosovo. This map will describe ERW hazard priority levels, while lending you the opportunity to demonstrate your cartographic expertise.

Layer or attribute	Description
Districts	**Kosovo third administrative layer boundaries polygons (feature class)**
NAME	District name
NAME_ALB	District name in Albanian
NAME_SERB	District name in Serbian
PREWARPOP	Prewar population
POSTWARPOP	Postwar population
MUNICIP	Municipality name
AREA	Area (square meters)
Kosovo	**Kosovo first administrative layer boundaries polygons (feature class)**
Municipalities	**Kosovo second administrative layer boundaries polygons (feature class)**
NAME	Municipality name
Towns	**Kosovo populated places points (feature class)**
NAME	Town name
NAME_ALB	Town name in Albanian
NAME_SERB	Town name in Serbian
MUNICIP	Municipality name
ClusterBombSites	**Kosovo cluster bomb-contaminated areas points (feature class)**
UXOLandmineSites	**Kosovo UXO and land-mine-contaminated areas polygons (feature class)**
Farmland	**Kosovo agricultural areas polygons (feature class)**
Woodland	**Kosovo forested and/or densely vegetated areas polygons (feature class)**
Roads	**Kosovo roads lines (feature class)**

Table 6.2   Data dictionary

Exercise 6.1

Transforming data using ArcToolbox

In this exercise, you will use a sequence of data "transformations" to calculate the total area contaminated by ERW hazards and then calculate the relative level of contamination for each district. Note that in the context of this chapter, "data transformation" means regenerating or re-creating already existing data so that it can be used to perform an analysis.

Create a buffer using proximity analysis

One of the most common categories involving such data transformation is called proximity analysis. ArcToolbox provides a Proximity Analysis toolset that includes the ability to create an area of interest within a specified distance around any point, line, or polygon feature. With this process, called "buffer analysis," you can create single- or multiple-ring buffers around features depending on application requirements.

You will now use buffer analysis to create hazard zones around all areas suspected of being contaminated by ERW.

1. **In ArcMap, open GISHUM_C6E1.mxd.**

A map symbolizing the locations of cluster bombs, UXOs, and land mines in Kosovo appears.

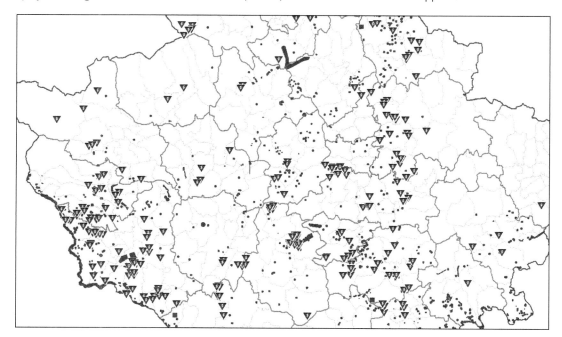

2. **Click the Toolbox icon to activate ArcToolbox.**

3. **In the ArcToolbox tool tree, expand Analysis Tools, then select Proximity > Buffer.**

4. **Populate the Buffer tool as follows:**

- Input Features: ClusterBombSites.
- Output Feature Class: ClusterBombSites_Buffer (add to your ERWContaminatedAreas feature dataset).
- Distance: Linear Units of 500 meters.
- Dissolve Type: ALL.
- Accept all other defaults.

5. **When your settings appear as follows, click OK.**

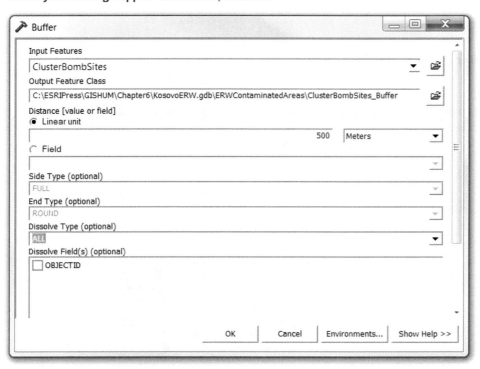

6. **When the buffer has been generated, close the operation box.**

The new layer is automatically added to your map display and table of contents.

Your turn

Create the following additional buffers:

- Towns_Buffer: 500 meters (save to your Administrative feature dataset)
- UXOLandmineSites_Buffer: 100 meters (save to your ERWContaminatedAreas feature dataset)

Why use a 500-meter buffer around cluster bombs and a 100-meter buffer around unexploded ordnance and land mines? Such determinations are based on the nature of the hazards, the accuracy of their locations, and other factors, all of which are variables and depend upon the conditions on the ground.

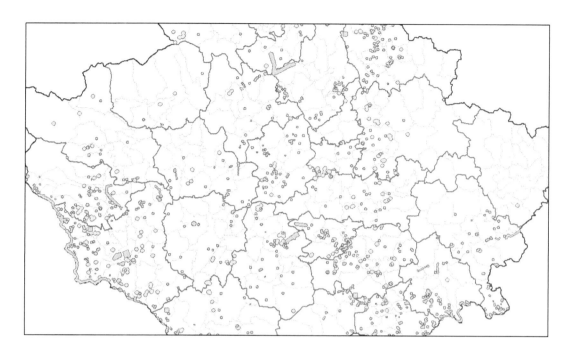

Your two buffered layers provide decision makers with a better appreciation of dangerous areas than the original dataset. You will now combine the two buffered layers into one consolidated layer.

Create a union using overlay analysis

Another very common category of data transformation is overlay analysis. The Overlay toolset in ArcToolbox includes a Union tool that creates a new feature class by combining the properties of two or more input layers. You will now use the Union tool to combine ClusterBombSites and UXOLandmineSites into one consolidated layer.

1. **Return to ArcToolbox. Go to Analysis Tools > Overlay > Union.**

2. **Populate the Union wizard as follows:**

    - Input Features: ClusterBombSites_Buffer and UXOLandmineSites_Buffer.
    - Output Feature Class: ERWBuffer_Union (add to your ERWContaminatedAreas feature dataset).
    - Join Attributes: ALL.
    - Accept all other defaults.

3. **When your settings appear as follows, click OK.**

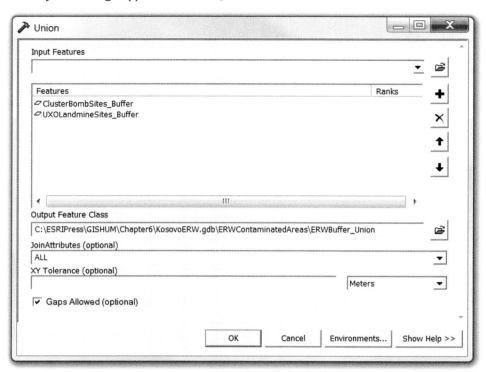

4. **Be patient. Complex data transformations like this one can take a few minutes even with a fast PC. When the union has been generated, close the operation box.**

You now have all the ERW contaminated areas in one spatial layer.

5. **Open the ERWBuffer_Union layer's attribute table.**

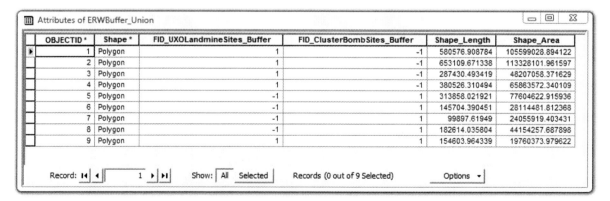

The union includes multiple records, created because the buffered layers contained both unique and overlapping features. To avoid double-counting the features, you will now aggregate the ERWBuffer_Union's features into one single feature.

Dissolve features using the Generalization tool

1. Return to ArcToolbox, select Data Management tools > Generalization > Dissolve.

2. Populate the Dissolve wizard as follows:

    *   Input Features: ERWBuffer_Union.
    *   Output Feature Class: ERWBuffer_Dissolve (add to your ERWContaminatedAreas feature dataset).
    *   Accept all other defaults.

3. When your settings appear as follows, click OK.

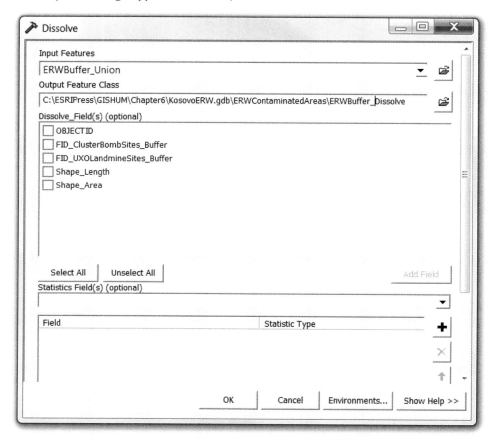

4. When the dissolve has been completed, close the operation box, and then open the new layer's attribute table.

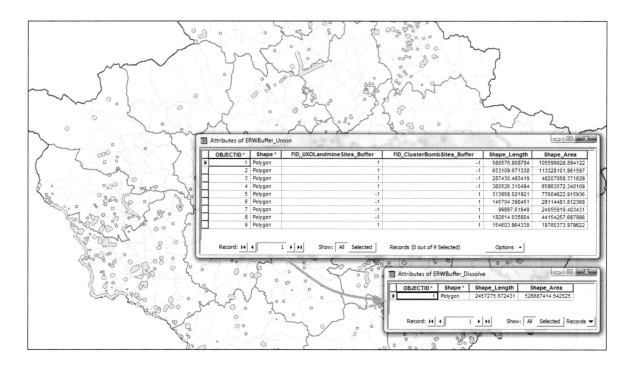

The dissolved layer's attribute table shows that all of the original layer's features have been aggregated into a single polygon representing ERW-contaminated areas throughout Kosovo.

You are now ready to analyze this ERWBuffer_Dissolve layer by district in order to determine the area of each district and the percentage of it that is contaminated by ERW.

Clip features using the Clip tool

ArcToolbox provides a set of tools that enable you to extract a new subset of features from an existing layer based upon certain selection criteria. Because we need to apportion the ERW-contaminated areas by district, you will now use a tool that will "clip" the Districts layer using your ERW-contaminated areas polygon.

1. **Return to ArcToolbox.**

2. **Choose Analysis Tools > Extract > Clip.**

3. **Populate the Clip wizard as follows:**

    - Input Features: Districts.
    - Clip Features: ERWBuffer_Dissolve.
    - Output Feature Class: ContaminatedDistricts (add to your ERWContaminatedAreas feature dataset, not the default Administrative feature dataset).
    - Accept all other defaults.

4. **When your settings appear as follows, click OK.**

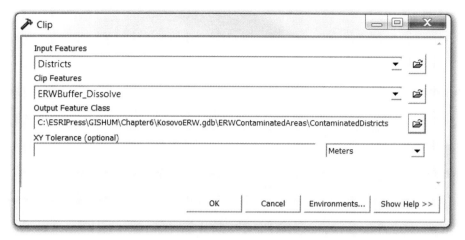

5. **When the clip is completed, close the operation box, and then open the new layer's attribute table.**

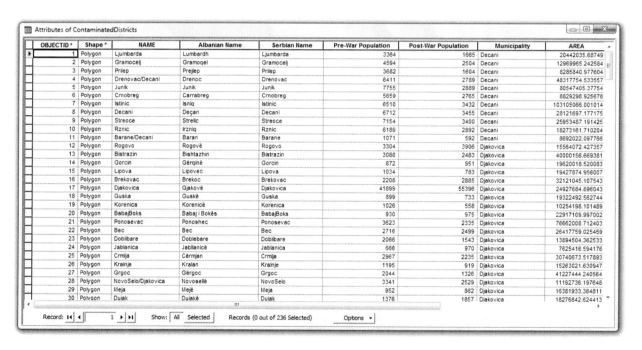

OBJECTID *	Shape *	NAME	Albanian Name	Serbian Name	Pre-War Population	Post-War Population	Municipality	AREA
1	Polygon	Ljumbarda	Lumbardh	Ljumbarda	3364	1685	Decani	20442035.68749
2	Polygon	Gramocelj	Gramoqel	Gramocelj	4594	2504	Decani	12969965.242584
3	Polygon	Prilep	Prejlep	Prilep	3682	1604	Decani	8285840.977604
4	Polygon	Drenovac/Decani	Drenoc	Drenovac	6411	2789	Decani	48317754.533557
5	Polygon	Junik	Junik	Junik	7755	2889	Decani	80547405.37754
6	Polygon	Crnobreg	Carrabreg	Crnobreg	5659	2765	Decani	8829298.925678
7	Polygon	Istinic	Isniq	Istinic	6510	3432	Decani	103105066.001014
8	Polygon	Decani	Deçan	Decani	6712	3455	Decani	28121697.177175
9	Polygon	Streoce	Strellc	Streoce	7154	3400	Decani	25953487.191425
10	Polygon	Rznic	Irzniq	Rznic	6189	2892	Decani	18273161.710204
11	Polygon	Barane/Decani	Baran	Barane	1071	592	Decani	8892022.097766
12	Polygon	Rogovo	Rogovë	Rogovo	3304	3906	Djakovica	15564072.427357
13	Polygon	Bistrazin	Bishtazhin	Bistrazin	3080	2483	Djakovica	40000156.669381
14	Polygon	Gorcin	Gërqinë	Gorcin	872	951	Djakovica	19620018.520083
15	Polygon	Lipova	Lipovec	Lipova	1034	783	Djakovica	19427874.956007
16	Polygon	Brekovac	Brekoc	Brekovac	2208	2885	Djakovica	32121045.107543
17	Polygon	Djakovica	Gjakovë	Djakovica	41899	55396	Djakovica	24927684.896043
18	Polygon	Guska	Guskë	Guska	899	733	Djakovica	19322492.562744
19	Polygon	Korenica	Korenicë	Korenica	1026	558	Djakovica	10254198.101489
20	Polygon	BabajBoks	Babaj i Bokës	BabajBoks	930	975	Djakovica	22917109.997002
21	Polygon	Ponosevac	Ponoshec	Ponosevac	3623	2335	Djakovica	76662008.712403
22	Polygon	Bec	Bec	Bec	2716	2499	Djakovica	26417759.025459
23	Polygon	Doblibare	Doblibare	Doblibare	2066	1543	Djakovica	13894504.362533
24	Polygon	Jablanica	Jabllanicë	Jablanica	666	970	Djakovica	7625416.594176
25	Polygon	Crmlja	Cërmjan	Crmlja	2967	2235	Djakovica	30740873.517893
26	Polygon	Kralnje	Kralan	Kralnje	1195	919	Djakovica	15263021.630947
27	Polygon	Grgoc	Gërgoc	Grgoc	2044	1326	Djakovica	41227444.240564
28	Polygon	NovoSelo/Djakovica	Novosellë	NovoSelo	3341	2529	Djakovica	11192736.197648
29	Polygon	Meja	Mejë	Meja	952	862	Djakovica	16381933.384811
30	Polygon	Duiak	Duiakë	Duiak	1376	1857	Djakovica	18276842.624413

Record: ◄ ◄ 1 ► ►| Show: All | Selected   Records (0 out of 236 Selected)   Options ▾

The table reveals that 236 of the original 327 districts in Kosovo have some level of ERW contamination. The map display shows the intersection of the input and clip layers, but it preserves the original attributes of the input features that were clipped. In other words, unlike the Intersect and Union geoprocessing operations, the attributes of the two inputs are not combined when using the Clip tool.

The field column AREA, therefore, reflects the total area of each contaminated district, not the specific area that is contaminated.

Exercise 6.2

Working with tabular data

So far, you have calculated the total area of Kosovo that was contaminated by ERW hazards. However, you can only prioritize ERW response measures based upon the relative (not absolute) level of contamination by district. In this exercise, you will calculate the percentage of contamination per district by summarizing tabular data values.

Create a new data field

1. Be sure that the ContaminatedDistricts attribute table is open.

2. Click the Options drop-down in the bottom right-hand corner of the attribute table window.

3. On the Options menu, choose Add Field.

4. Populate the Add Field dialog box as follows:

    *   Name: AREACONTM.
    *   Type: Double.
    *   Alias: Area Contaminated.
    *   Accept all other defaults.

5. When your settings appear as follows, click OK.

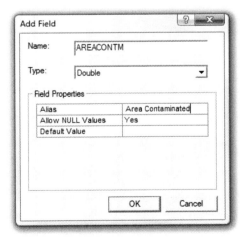

The attribute table now contains a new column in which you will calculate what percentage of each district is contaminated.

Calculate feature geometry

Whenever you are working with an attribute table of a feature layer, you can easily calculate the area, perimeter, length, x-location, y-location, x-centroid, or y-centroid using the Calculate Geometry function in ArcMap.

Geometry calculations are planimetric. What that means for us here is that you can calculate the area, length, or perimeter of features only if the coordinate system being used is a projected coordinate system. If the data source uses a geographic coordinate system instead, such as WGS 1984, you must apply the projected coordinate system of the data frame, or manually project the data source, before performing the calculations.

1. **Right-click the new field's header, and then select Calculate Geometry. Click Yes if asked whether you wish to calculate outside of an edit session.**

2. **The default settings should appear as in the following screen capture.**

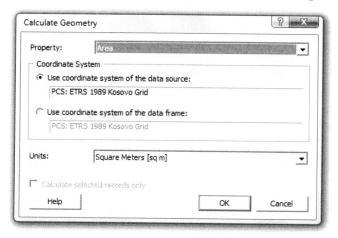

3. **Click OK.**

Your table now includes both the total area of each district and the area of each district that is contaminated with ERW. Now you can determine the percentage of each district that is contaminated.

4. **On the Options menu, choose Add Field.**

5. **Populate the Add Field dialog box as follows:**

    *   Name: PCTCONTM.
    *   Type: Float.
    *   Alias: Percent Contaminated.
    *   Accept all other defaults.

6. When your settings appear as follows, click OK.

Your attribute table now contains a new column in which you will calculate the percentage of contamination for each district.

7. **To populate the new field, right-click the field heading, and then select Field Calculator.**

8. **In the Field Calculator dialog box, build the following expression:**
    [AREACONTM] /[AREA]*100

9. **When your dialog box appears as follows, click OK.**

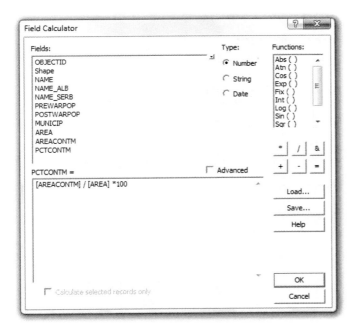

The ContaminatedDistricts layer is ready to be analyzed.

Sort fields by attribute value

1. To determine which district has the most contaminated land, right-click the Percent Contaminated field, and then sort the districts in descending order.

2. Right-click the following fields, and then select Turn Field Off:

    *   Albanian Name
    *   Serbian Name
    *   Pre-War Population
    *   Post-War Population
    *   Shape_Length
    *   Shape_Area

You can always go to Options > Turn All Fields On whenever you need to restore your table, but for now it will be easier to limit the table to just a few fields.

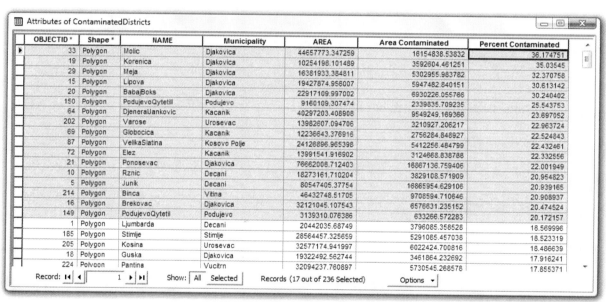

Notice that 17 districts have contamination levels greater than 20 percent, with Molic being the most affected. Additionally, the five worst-contaminated districts are all within the municipality of Djakovica. Clearly this is the most severely contaminated region of Kosovo and should be a top priority for the ERW response teams.

The highlighted records in the preceding figure were manually selected by clicking the gray bar to the left of each feature while holding the Ctrl key on the keyboard. You will soon learn how to more efficiently select features by attribute values.

You will now rank each district's level of contamination using table 6.1 on page 189.

3. **On the Options menu, choose Add Field.**

4. **Populate the Add Field dialog box as follows:**

    *   Name: LVLCONTM.
    *   Type: Text.
    *   Alias: Level Contaminated.
    *   Length: 10.
    *   Accept all other defaults.

5. **When your settings appear as follows, click OK.**

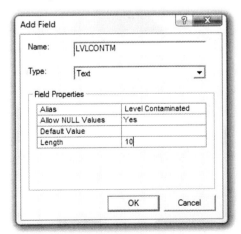

You are now ready to filter your attribute table according to the criteria for low, medium, and high levels of contamination.

6. **Within the ContaminatedDistricts attribute table, click Options and then Select By Attribute.**

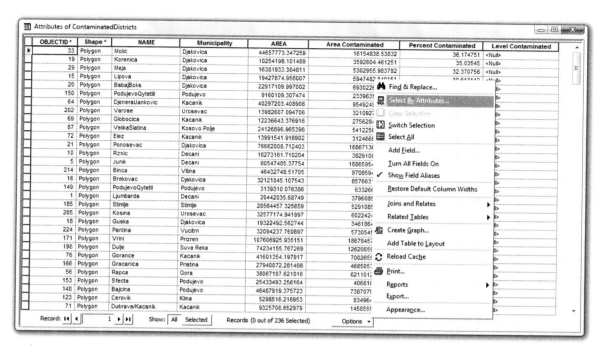

OBJECTID *	Shape *	NAME	Municipality	AREA	Area Contaminated	Percent Contaminated	Level Contaminated
33	Polygon	Molic	Djakovica	44657773.347259	16154838.53832	36.174751	<Null>
19	Polygon	Korenica	Djakovica	10254198.101489	3592604.461251	35.03545	<Null>
29	Polygon	Meja	Djakovica	16381933.384611	5302955.983782	32.370758	<Null>
15	Polygon	Lipova	Djakovica	19427874.956007	5947482.949454		
20	Polygon	BabajBoks	Djakovica	22917109.997002	6930226		
150	Polygon	PodujevoQytetill	Podujevo	9160109.307474	2339635		
64	Polygon	DjeneraUankovic	Kacanik	40297203.408908	9549245		
202	Polygon	Varose	Urosevac	13982607.094706	3210927		
69	Polygon	Globocica	Kacanik	12236643.376916	2756286		
87	Polygon	VelikaSlatina	Kosovo Polje	24126896.965398	5412256		
72	Polygon	Elez	Kacanik	13991541.916902	3124668		
21	Polygon	Ponosevac	Djakovica	76662006.712403	16867136		
10	Polygon	Rznic	Decani	18273161.710204	3829108		
5	Polygon	Junik	Decani	80547405.37754	16865954		
214	Polygon	Binca	Vitina	48432748.51705	9708594		
16	Polygon	Brekovac	Djakovica	32121045.107543	657663		
149	Polygon	PodujevoQytetil	Podujevo	3139310.076386	633266		
1	Polygon	Ljumbarda	Decani	20442035.68749	3796085		
185	Polygon	Stimlje	Stimje	28564457.325659	5291085		
205	Polygon	Kosina	Urosevac	32577174.941997	6022424		
18	Polygon	Guska	Djakovica	19322492.562744	3461884		
224	Polygon	Pantina	Vucitrn	32094237.760897	5730545		
171	Polygon	Vrini	Prizren	107606925.935151	18678457		
198	Polygon	Dulje	Suva Reka	74234155.767269	12620655		
76	Polygon	Gorance	Kacanik	41601354.197917	7002655		
166	Polygon	Gracanica	Pristina	27940072.261408	4685052		
56	Polygon	Rapca	Gora	38067107.621016	6211017		
153	Polygon	Sfecta	Podujevo	25433493.256164	406818		
148	Polygon	Bajcina	Podujevo	46487919.375723	7387079		
123	Polygon	Cerovik	Klina	5298516.216953	83496		
71	Polygon	Dubrava/Kacanik	Kacanik	9325708.652979	1458555		

Menu items shown: Find & Replace..., Select By Attributes..., Clear Selection, Switch Selection, Select All, Add Field..., Turn All Fields On, ✓ Show Field Aliases, Restore Default Column Widths, Joins and Relates ▶, Related Tables ▶, Create Graph..., Add Table to Layout, Reload Cache, Print..., Reports ▶, Export..., Appearance...

Record: |◄ ◄ | 1 | ► ►| Show: All Selected   Records (0 out of 236 Selected)   Options ▼

7. In the Select By Attributes dialog box, double-click "PCTCONTM" in the field box and build the following expression:
"PCTCONTM"<1.3

8. When your settings appear as follows, click Apply.

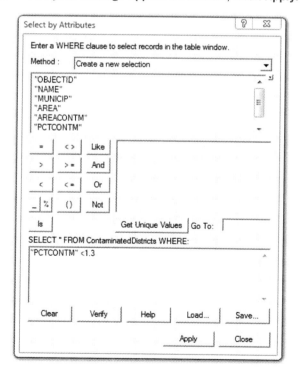

All districts with less than 1.3 percent contamination will be selected in the attribute table. You will now assign those selected districts the contamination level of Low.

9. Right-click the Level Contaminated header, and then select Field Calculator.

10. Populate the Field Calculator dialog box as follows:

    - Type: String.
    - LVLCONTM: "Low".
    - Accept all other defaults.

11. When your settings appear as follows, click **OK**.

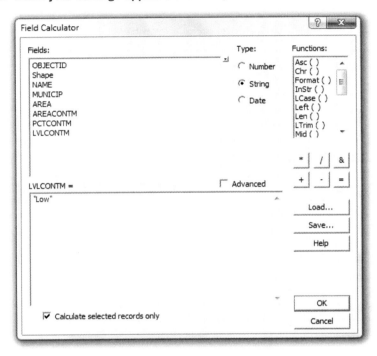

The selected records have now been assigned the contamination ranking of Low.

12. Using the same process, populate the rankings of those districts falling into the Medium and High levels of contamination. Use the following expressions to select your districts:

    Medium   **"PCTCONTM" >=1.3 AND "PCTCONTM" <= 6.5**

    High   **"PCTCONTM" > 6.5**

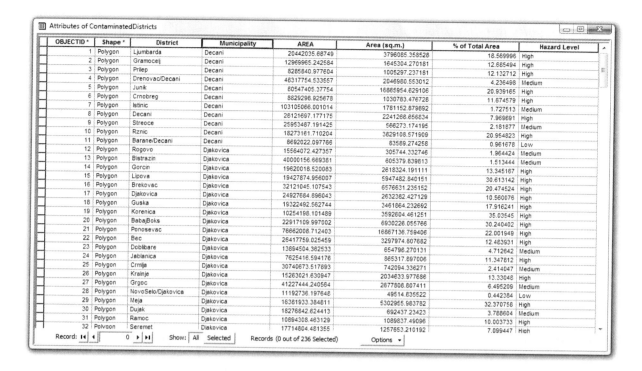

OBJECTID *	Shape *	District	Municipality	AREA	Area (sq.m.)	% of Total Area	Hazard Level
1	Polygon	Ljumbarda	Decani	20442035.68749	3796085.358528	18.569996	High
2	Polygon	Gramocelj	Decani	12969965.242584	1645304.270181	12.685494	High
3	Polygon	Prilep	Decani	8285840.977604	1005297.237181	12.132712	High
4	Polygon	Drenovac/Decani	Decani	48317754.533557	2046980.553012	4.236498	Medium
5	Polygon	Junik	Decani	80547405.37754	16865954.629106	20.939165	High
6	Polygon	Crnobreg	Decani	8829296.925678	1030783.476728	11.674579	High
7	Polygon	Istinic	Decani	103105066.001014	1781152.879892	1.727513	Medium
8	Polygon	Decani	Decani	28121697.177175	2241266.656834	7.969691	High
9	Polygon	Streoce	Decani	25953467.191425	566273.174195	2.181877	Medium
10	Polygon	Rznic	Decani	18273161.710204	3829108.571909	20.954823	High
11	Polygon	Barane/Decani	Decani	8692022.097766	83589.274258	0.961678	Low
12	Polygon	Rogovo	Djakovica	15564072.427357	305744.332746	1.964424	Medium
13	Polygon	Bistrazin	Djakovica	40000156.669381	605379.839813	1.513444	Medium
14	Polygon	Gorcin	Djakovica	19620018.520083	2618324.191111	13.345167	High
15	Polygon	Lipova	Djakovica	19427874.956007	5947482.840151	30.613142	High
16	Polygon	Brekovac	Djakovica	32121045.107543	6576631.235152	20.474524	High
17	Polygon	Djakovica	Djakovica	24927684.896043	2632382.427129	10.560076	High
18	Polygon	Guska	Djakovica	19322492.562744	3461864.232692	17.916241	High
19	Polygon	Korenica	Djakovica	10254198.101489	3592604.461251	35.03545	High
20	Polygon	BabajBoks	Djakovica	22917109.997002	6930226.055766	30.240402	High
21	Polygon	Ponosevac	Djakovica	76662008.712403	16867136.759406	22.001949	High
22	Polygon	Bec	Djakovica	26417759.025459	3297974.807682	12.483931	High
23	Polygon	Doblibare	Djakovica	13894504.362533	654796.270131	4.712642	Medium
24	Polygon	Jablanica	Djakovica	7625416.594176	865317.897006	11.347812	High
25	Polygon	Crmlja	Djakovica	30740673.517893	742094.336271	2.414047	Medium
26	Polygon	Krainje	Djakovica	15263021.630947	2034633.977686	13.33048	High
27	Polygon	Grgoc	Djakovica	41227444.240564	2677808.807411	6.495209	Medium
28	Polygon	NovoSelo/Djakovica	Djakovica	11192736.197648	49514.835522	0.442384	Low
29	Polygon	Meja	Djakovica	16381933.384811	5302955.983782	32.370758	High
30	Polygon	Dujak	Djakovica	18276842.624413	692437.23423	3.788604	Medium
31	Polygon	Ramoc	Djakovica	10894308.463129	1089837.49096	10.003733	High
32	Polygon	Seremet	Djakovica	17714604.481355	1257653.210192	7.099447	High

Record: ◄◄ ◄   0   ► ►◄   Show: All Selected   Records (0 out of 236 Selected)   Options ▾

Your table now ranks each district according to the definition of low, medium, and high levels of ERW contamination. This simple table allows for immediate insight into a very complex situation and provides planners with systematic, actionable information.

Your turn

Using the methods learned in chapter 2, join the ContaminatedDistricts layer to the Districts layer and create a choropleth map showing the level of ERW contamination of Kosovo's districts.

The simple example below provides a baseline for your work. Use this opportunity to test your ability to create effective thematic maps and try to improve upon the example.

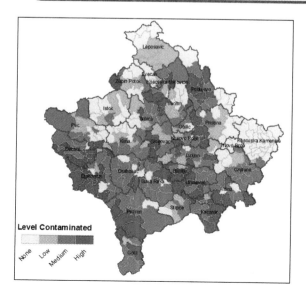

Export results in tabular format

GIS specialists export spatially generated data tables so that the information can be analyzed statistically or shared with others working outside a GIS environment.

1. Clear any ContaminatedDistricts features that are selected. (You can do this in one of three easy ways: 1. From within the attribute table, go to Options > Clear Selection; 2. On the main menu, go to Selection > Clear Selected Features; or 3. On the Tools toolbar, click the Clear Selected Features ⊠ button.)

2. Now from within the attribute table, go to Options > Export.

3. Export the table to the Chapter6 folder as **ContaminatedAreasByDistrict.dbf**. Click No if asked whether to add the new table to the current map.

What to turn in

If you are working in a classroom setting with an instructor, submit an electronic or printed version of the ContaminatedAreasByDistrict.dbf and your choropleth map.

Exercise 6.3

Transforming data using ArcToolbox (additional examples)

One of the first steps in a humanitarian crisis is providing direct support to facilitate the movement of people and supplies. In this exercise, you will perform more analysis with your Kosovo geodatabase in order to create 200-meter buffers around roads and then intersect that buffered layer with the two ERW-hazards files. This will isolate areas of greatest human vulnerability.

Your objective in this exercise is to develop guidance for ERW response teams in deciding whether to clear or close off contaminated areas. Just because an area is highly contaminated does not necessarily mean that it will be cleared—response teams may decide to simply close the area so that they can devote their resources to other areas that are more essential to human passage. As a rule, your role as a humanitarian GIS professional should always be to provide decisive information without prejudicing the decision-making process.

Create a buffer using proximity analysis

1.   **In ArcMap, open GISHUM_C6E3.mxd.**

2.   **In the ArcToolbox tool tree, expand Analysis Tools, then select Proximity > Buffer.**

3.   **Populate the Field Calculator dialog box as follows:**

   • Input Features: Roads.
   • Output Feature Class: Roads_Buffer (add directly to your geodatabase, or use ArcCatalog to create a Transportation feature dataset for both the original and buffered roads layers).
   • Distance Linear Unit: 200 meters.
   • Dissolve Type: NONE.
   • Accept all other defaults.

4. When your settings appear as follows, click OK.

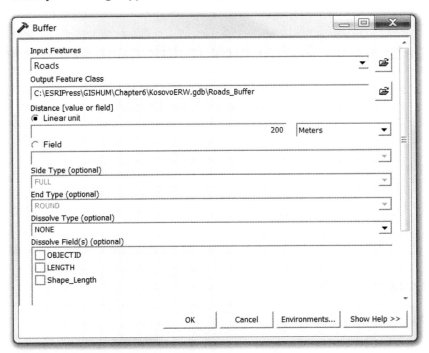

5. When the buffer is completed, close the operation box.

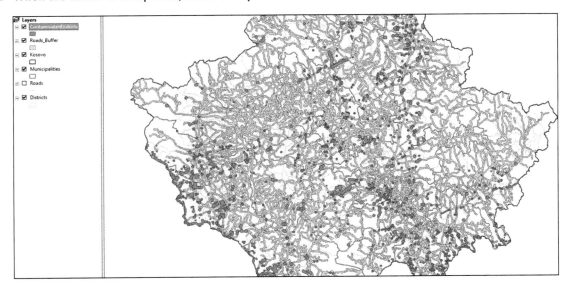

You may need to adjust your layer settings to replicate the image.

Create an intersect using overlay analysis

The Overlay toolset in ArcToolbox includes an Intersect tool that creates a new feature class from overlapping the properties of two or more input layers. You will now use the Intersect tool to identify those sections of the road buffer that overlap with areas that are contaminated with ERW.

1. **Return to ArcToolbox. Go to Analysis Tools > Overlay > Intersect.**

2. **Populate the Field Calculator dialog box as follows:**

   - Input Features: Roads_Buffer and ContaminatedDistricts.
   - Output Feature Class: ContaminatedRoads (add to your ERWContaminatedAreas feature dataset).
   - Accept all other defaults.

3. **When your settings appear as follows, click OK.**

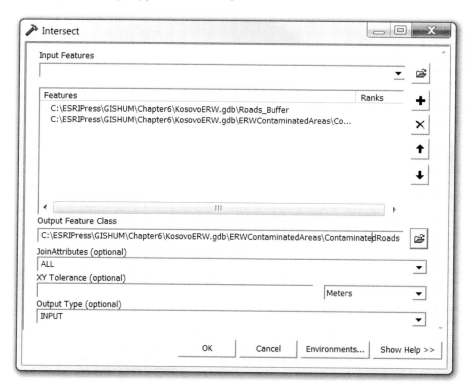

4. **When the analysis is complete, close the operation box, and then open the new layer's attribute table.**

Each contaminated road segment is identified as a separate record, enabling you to create a report containing maps and tables to guide humanitarian planners in deciding the best action to take—for example, what type of signage to post or where to locate those signs.

To make our analysis simpler, we will now dissolve the features into one single feature within the ContaminatedRoads layer.

5. **Return to ArcToolbox, and select Data Management tools > Generalization > Dissolve.**

6. **Populate the Dissolve wizard as follows:**

- Input Features: ContaminatedRoads.
- Output Feature Class: ContaminatedRoads_Dissolve (add to your ERWContaminatedAreas feature dataset).
- Accept all other defaults.

7. **When your settings appear as follows, click OK.**

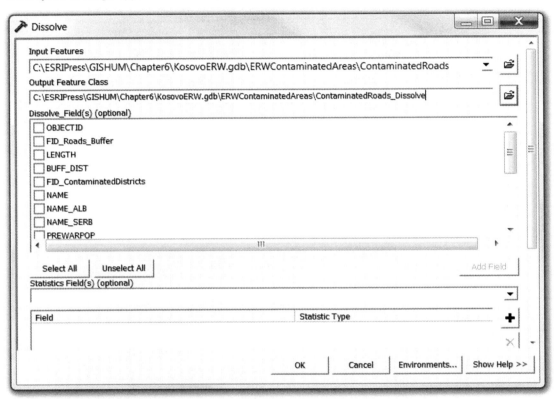

8. **When the dissolve has been completed, close the operation box, and then open the new layer's attribute table.**

Note that the features are now consolidated into one single polygon. Your map display may look the same, but closer examination reveals that by generalizing the ContaminatedRoads layer you have significantly simplified your active dataset. This not only makes it aesthetically more pleasing, but it also reduces the file size of the layer.

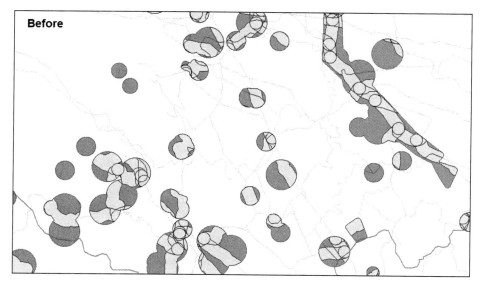

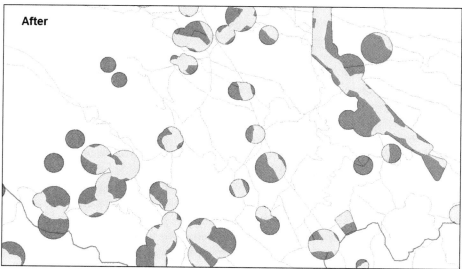

Adjust your layer settings as appropriate.

Exercise 6.4

Working with tabular data (additional examples)

You are now ready to prioritize the decontamination of hazardous areas based upon the following criteria:

Argument	Points added
Contaminated area intersects with town buffer	10
Contaminated area intersects with road buffer	5
Contaminated area intersects with farmland	2
Contaminated area intersects with woodland foraging area	1.5
Maximum hazard points	18.5

Table 6.3 Hazard weighting system

Priority	Hazard points
Low	< 3.5
Medium	3.5–9
High	>9(max = 18)

Table 6.4 Priority levels

The MACC staff selected these criteria after careful consideration of available ERW clearance capacity and the need to be unbiased and transparent in determining hazard reduction strategies. In your role as a humanitarian GIS professional, always try to obtain "reality checks" like this so that your analysis can be grounded by the resource constraints confronted at the field level.

Create new types of data fields

You are now ready to assign hazard points to each contaminated area according to the MACC's criteria.

1. **In ArcMap, open GISHUM_C6E4.mxd**

The map display is an extremely busy workspace and contains original ERW layers as well as the buffers you created. It also contains polygons delineating farmland and woodland, in accordance with the hazard weighting criteria.

2. **Open the ClusterBombSites attribute table, and then go to Options > Add Field.**

3. **Add a new field as follows:**

    *   Name: HZDPTS.
    *   Type: Float.
    *   Alias: Hazard Points.
    *   Accept all other defaults.

4. **When your settings appear as follows, click OK.**

5. Add another new field as follows:

   - Name: PRIORITY.
   - Type: Text.
   - Alias: Hazard Priority.
   - Length: 10.
   - Accept all other defaults.

6. When your settings appear as follows, click OK.

7. Add the same two new fields in the UXOLandmineSites layer.

Select features by location

1. On the main menu, go to Selection > Select By Location.

2. Populate the dialog box to select features from the ClusterBombSites and UXOLandmineSites layer that intersect the features in the Towns_Buffer layer.

3. **When your settings appear as follows, click OK.**

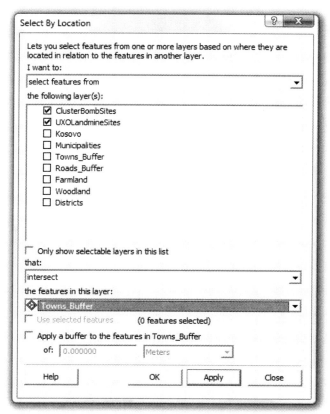

You have now selected any ERW sites within 500 meters of towns. Note that even if you did not make a Towns_Buffer layer earlier in this chapter, you can still perform this operation by applying a buffer within the Select By Location dialog box.

4. **Open the attribute table for the Cluster bomb layer.**

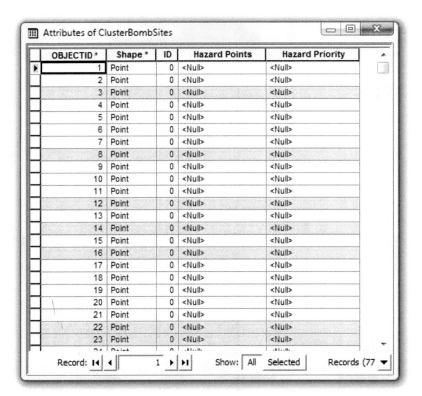

77 of the 518 cluster bomb sites are selected.

5. **Right-click the Hazard Points header, and then open the Field Calculator.**

6. **Assign the selected records 10 points, as per the MACC's hazard weighting system.**

7. **When your settings appear as follows, click OK.**

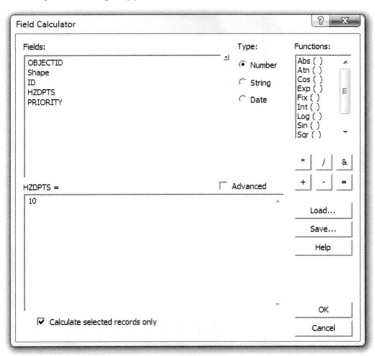

8. Using the same method, assign 10 hazard points to the UXOLandmineSites layer.

9. Close the open attribute tables, and then clear the selected features 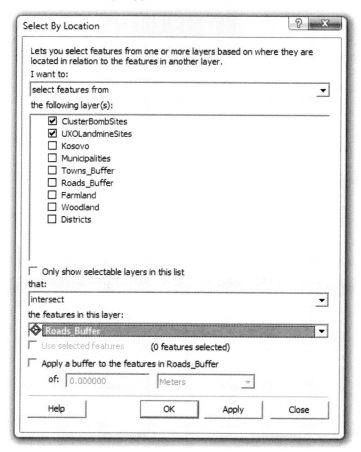 .

10. On the main menu, go to Selection > Select By Location.

11. Populate the dialog box to select features from the ClusterBombSites and UXOLandmineSites layer that intersect the features in the Roads_Buffer layer.

12. When your settings appear as follows, click OK.

13. Open the attribute table for the ClusterBombSites layer.

Note that you have already identified some of the selected sites as also being within 500 meters of a town. You will now assign additional hazard points to the selected features since they are within 200 meters of a road.

14. Right-click the Hazard Points header, and then open the Field Calculator.

15. Build the following expression in order to add 5 points to the selected cluster bomb sites, as per the MACC's hazard weighting system: HZDPTS=[HZDPTS]+5

16. When your settings appear as follows, click OK.

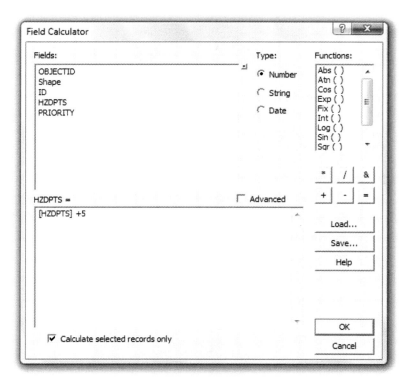

17. **Using the same method, add 5 hazard points for the selected UXO and land mine areas.**

18. **Close the open attribute tables, and then clear the selected features**

Your turn

You are getting closer to producing a weighted analysis of ERW contamination in Kosovo. Complete the hazard weighting process introduced earlier in this chapter. Select all ERW sites that intersect farmland and assign them 2 hazard points. Then select all ERW sites that intersect woodland and assign them 1.5 hazard points. Remember to clear selected features in between and after finishing your hazard weighting. You will soon be able to produce a map that illustrates these hazard priority levels, empowering humanitarian planners to easily decipher your analysis and determine optimal response strategies more quickly.

Select features by attribute

Now that hazard points have been assigned to the ERW sites, you can now associate each site's weighting with the priority levels determined by the MACC.

Priority	Hazard Points
Low	< 3.5
Medium	3.5–9
High	>9

Table 6.5

To do this you will select ERW sites according to the value of their attribute, Hazard Points.

1. Open the attribute table for the ClusterBombSites layer.

2. Go to Options > Select by Attributes. (Alternatively, you can use the main menu's Selection > Select by Attributes.)

3. Build the following expression in order to select features from the ClusterBombSites with less than 3.5 hazard points: "HZDPTS" IS NULL OR "HZDPTS"<3.5.

4. When your settings appear as follows, click Apply.

5. Right-click the Hazard Points header, and then open the Field Calculator.

6. Assign the selected features with a hazard priority level of Low, as per the MACC's hazard weighting system.

7. **When your settings appear as follows, click OK.**

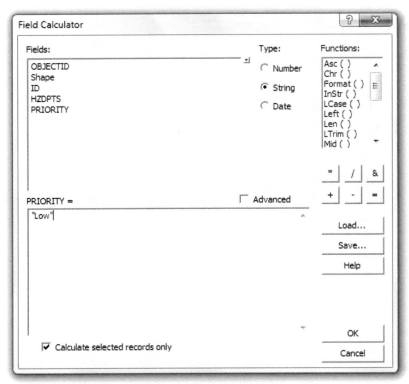

All ClusterBombSites with less than 3.5 hazard points, or null values, have now been assigned a Low hazard priority level.

8. **Using the same method, assign all UXO and land mine areas with less than 3.5 hazard points a Low hazard priority level.**

9. **Clear the selected features.**

Your turn

Complete the hazard prioritizing process for ERW sites with medium and high priority levels. Use the following expressions to select by attribute value:

Medium   "HZDPTS">= 3.5AND"HZDPTS"<=9
High   "HZDPTS">9

Remember to clear selected features in between and after finishing your hazard prioritization. The first few records of your completed attribute tables should look as follows. (You may want to make your tables more legible by turning off shape fields and assigning the numerical value of zero to <Null> hazard point values.)

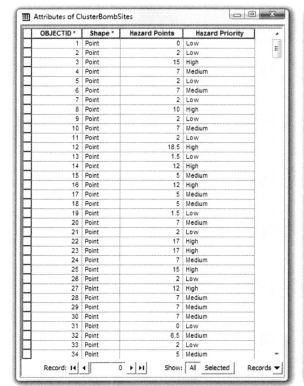

Attributes of ClusterBombSites

OBJECTID *	Shape *	Hazard Points	Hazard Priority
1	Point	0	Low
2	Point	2	Low
3	Point	15	High
4	Point	7	Medium
5	Point	2	Low
6	Point	7	Medium
7	Point	2	Low
8	Point	10	High
9	Point	2	Low
10	Point	7	Medium
11	Point	2	Low
12	Point	18.5	High
13	Point	1.5	Low
14	Point	12	High
15	Point	5	Medium
16	Point	12	High
17	Point	5	Medium
18	Point	5	Medium
19	Point	1.5	Low
20	Point	7	Medium
21	Point	2	Low
22	Point	17	High
23	Point	17	High
24	Point	7	Medium
25	Point	15	High
26	Point	2	Low
27	Point	12	High
28	Point	7	Medium
29	Point	7	Medium
30	Point	7	Medium
31	Point	0	Low
32	Point	6.5	Medium
33	Point	2	Low
34	Point	5	Medium

Record: 0   Show: All Selected   Records

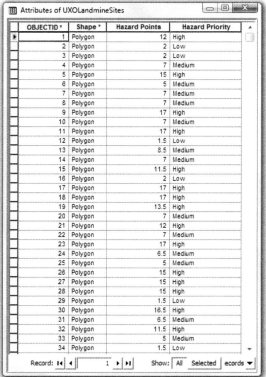

Attributes of UXOLandmineSites

OBJECTID *	Shape *	Hazard Points	Hazard Priority
1	Polygon	12	High
2	Polygon	2	Low
3	Polygon	2	Low
4	Polygon	7	Medium
5	Polygon	15	High
6	Polygon	5	Medium
7	Polygon	7	Medium
8	Polygon	7	Medium
9	Polygon	17	High
10	Polygon	7	Medium
11	Polygon	17	High
12	Polygon	1.5	Low
13	Polygon	8.5	Medium
14	Polygon	7	Medium
15	Polygon	11.5	High
16	Polygon	2	Low
17	Polygon	17	High
18	Polygon	17	High
19	Polygon	13.5	High
20	Polygon	7	Medium
21	Polygon	12	High
22	Polygon	7	Medium
23	Polygon	17	High
24	Polygon	6.5	Medium
25	Polygon	5	Medium
26	Polygon	15	High
27	Polygon	15	High
28	Polygon	15	High
29	Polygon	1.5	Low
30	Polygon	16.5	High
31	Polygon	6.5	Medium
32	Polygon	11.5	High
33	Polygon	5	Medium
34	Polygon	1.5	Low

Record: 1   Show: All Selected   ecords

Your ERW attribute tables reveal the range of ERW hazard values that can be extrapolated from just a few simple criteria. The hazard values can now be communicated to ERW clearance teams and other local partners in order to implement appropriate mitigation strategies.

Assignment

Prepare a thematic map illustrating the ERW hazard in the two municipalities of Djakovica and Decani in eastern Kosovo. The following figure demonstrates a complex cartographic layout that includes two data frames, a title, two legends, a scale bar, and varying symbols. Can you replicate it or improve upon it further?

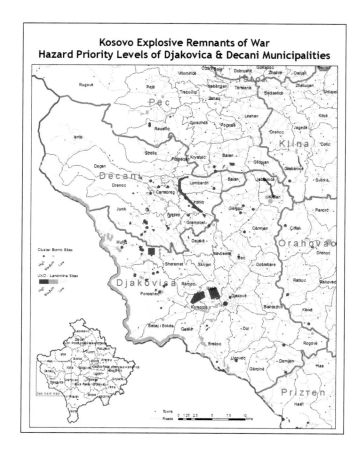

Kosovo Explosive Remnants of War
Hazard Priority Levels of Djakovica & Decani Municipalities

What to turn in

If you are working in a classroom setting with an instructor, submit an electronic or printed version of your thematic map.

Optional assignment

One of the most important skills of a humanitarian GIS specialist is the ability to present insights to both technical and nontechnical audiences in a way that they can both understand, allowing them to act upon those insights and thereby improve the overall success of the mission.

Prepare a strategy brief using additional map products to describe the ERW threat that existed in Kosovo and how it affected land use and navigation by local and international stakeholders. Analyze the correlation between prewar and postwar populations of each district and its level of ERW contamination. Be careful not to overstate your findings; qualify any assumptions that you made during your analysis.

What to turn in

If you are working in a classroom setting with an instructor, submit an electronic or printed version of your brief.

You have begun to discover the power of GIS-based analysis for promoting safe and effective humanitarian operations.

This chapter provides multiple examples of a specific category of analytical GIS: transformations. Your ability to "transform" (in other words, *utilize*) whatever data that is available in a humanitarian emergency into decisive information will make you a valuable member of the humanitarian community. Work hard at those skills, and develop confidence in using ArcGIS beyond its excellent cartographic applications.

Additional reading

International Committee of the Red Cross. 2001. Cluster bombs and landmines in Kosovo—Explosive remnants of war. Revised edition. Mines-Arms Unit – ICRC, June 2001, Geneva, available for download at http://www.icrc.org/Web/eng/siteeng0.nsf/html/explosive-remnants-of-war-brochure-311201.

Messick, Shawn. 2000. Kosovo (FRY) landmine impact survey. Report for the Survey Action Center, VVAF, March 2000, Washington, D.C., available online at http://www.mineaction.org/docs/1006648832_.asp.

Chapter 7

Planning logistics using Network Analyst

Delivering food, water, and other forms of assistance to disaster-affected populations is an urgent and continual requirement during humanitarian emergencies, with massive transport cost implications. Improving the efficiency of relief distribution has become an ongoing priority for donors and their implementing partners.

The UN World Food Programme (WFP) has studied how transportation network analysis, a powerful type of spatial analysis, can be used to support its operations in Ethiopia. This type of analysis requires data and complex models that rarely exist in the developing world. The WFP's longer-term missions sometimes offer enough time to implement such models, enabling logistics planners to solve difficult routing problems using ArcGIS-based network analysis.

In this lesson, you will use the UNSDI-T data model you studied in chapter 5 to perform network analysis to locate warehouses, plan delivery routes, and simulate the effect of route closures in an ongoing humanitarian emergency in Ethiopia.

Network analysis supports the World Food Programme

Rising fuel prices, the need to reduce carbon emissions, and (most important) the need to effectively serve widely dispersed populations, are some of the many reasons to improve the efficiency of humanitarian relief operations. Through its partnership with TNT (a global transportation and distribution company) and with technical support from the UN Joint Logistics Centre (UNJLC), the WFP has been investigating ways to optimize its supply chains and transportation networks.

The approach originated with an initiative led by UNJLC and WFP in 2004 to develop a route optimization tool for Sudan in collaboration with TNT. The original aim was to devise a way to quantify what transport savings might be gained from various road upgrading options. The tool was subsequently implemented in Cambodia to study the transport efficiency of WFP's distribution point configuration. Then, using an enhanced version of this tool, the methodology and lessons learned from this project were successfully applied to an analysis of WFP's School Feeding Programme's land transport network in Liberia. This showed that cost savings of up to 20 percent could be achieved through a reconfiguration of warehouse locations and through better convoy planning and dispatching.

The latest implementation of this methodology has been adopted by WFP Ethiopia to support its Targeted Supplementary Feeding Programme (TSFP). Considering TSFP's particular requirement of delivering small quantities of food to a large number of destinations, all within a strict time frame, planners quickly recognized the problem: a large number of low-carrying-capacity vehicles were making long and expensive trips around Ethiopia to service WFP's configuration of five centralized warehouses. They decided to focus the transport optimization analysis on quantifying the potential savings in travel time and cost to be gained from adding warehouses closer to destination clusters. The objective is to identify the most efficient configuration among a number of alternative proposals. Ultimately, this will allow for heavier trucks to pre-position larger quantities of food before final delivery by light vehicles. For overviews of the WFP's activities in Ethiopia, visit these Web sites: `http://www.wfp.org/countries/ethiopia` and `http://www.wfp.org/videos/horn-africa`.

Through a series of exercises in the use of ArcGIS Network Analyst, this chapter illustrates the variety of ways in which the tool supports logistics decision making. With it, you can solve common network problems, such as finding the best route across a city, the closest emergency vehicle or facility, and the service area around a location for servicing a set of orders with a fleet of vehicles.

Finding the best route. ArcGIS Network Analyst finds the best way to get from one place to another while visiting several locations along the way. Such locations can be specified interactively by placing points on the screen, entering an address, or using points in an existing feature class or feature layer. Network Analyst can determine the best route in an order of locations specified by the user or it can determine the best sequence itself.

Locating the closest facility. Finding the hospital closest to an accident, the police car nearest a crime scene, and the store most convenient to a customer are all examples of closest facility problems. Using the tool in search of nearby facilities, you can specify how many to locate and whether the direction of travel is toward or away from them. Having identified the closest facilities, you can display the best route to or from them, including directions to each facility, and return the travel cost for each route. Also, you can specify an impedance cutoff beyond which

ArcGIS Network Analyst should not search for a facility. For instance, with a closest facility problem set up to search for hospitals within 15 minutes' drive time of an accident site, any hospital requiring longer than 15 minutes to reach will not be included in the results.

Identifying service areas. With Network Analyst, you can find service areas around any location on a network. A network service area is a region that encompasses all accessible streets, that is, streets that lie within a specified impedance. For instance, the 10-minute service area for a facility includes all the streets that can be reached within 10 minutes from that facility.

Creating an OD cost matrix. Not surprisingly, the path of least resistance (lowest impedance) is also the least-cost path. An OD (origin-destination) cost matrix is a table that contains the network impedance from each origin to each destination. In ascending order, it ranks the destinations that each origin connects to, based on the minimum network impedance required to travel from that origin to each destination. With ArcGIS Network Analyst, you can create an OD cost matrix from multiple origins to multiple destinations.

The tool discovers the best network path for each origin-destination pair, and the cost is stored in the attribute table of the output lines, which are straight lines. Note that the term "cost" is not necessarily a monetary measure: you can use many different types of attribute values to minimize in this type of analysis, including travel time, vehicle efficiency, or route risk. Cost can refer to service quality, operating expense, safety levels, and so on.

Solving a vehicle routing problem. A dispatcher managing a fleet of vehicles is often required to make decisions about vehicle routing, such as how best to assign a group of customers to a fleet of vehicles and to sequence and schedule their visits. The objectives in solving such vehicle routing problems (VRP) are to provide a high level of customer service by honoring any time windows, while keeping the overall operating and investment costs for each route as low as possible. The constraints are to complete the routes with available resources and within the time limits imposed by driver work shifts, driving speeds, and customer commitments. The vehicle-routing problem solver within ArcGIS Network Analyst can determine solutions for such complex fleet management tasks.

Consult ArcGIS Help for a more complete review of Network Analyst's functionality and operation. For a video introduction to network analysis, see http://www.esri.com/software/arcgis/extensions/networkanalyst/demos.html.

UNSDI-T data model. Before beginning the exercises in this chapter, you may want to review your work in chapter 5 and visit the UNSDI-T Web site: http://www.logcluster.org/tools/mapcentre/unsdi/unsdi-t-v2.0. (If this link has expired, search the Internet for "UNSDI-T" and find the latest home page for the UNSDI for Transport data model.)

Scenario: Optimizing supply chains in Ethiopia

In this chapter, you will use the geodatabase introduced in chapter 5 to perform a series of logistics planning exercises for the WFP's operations in Ethiopia. In exercise 7.1, you will build and visualize a network dataset and set impedance values. In the next exercise, you generate an origin-destination matrix and calculate distances and travel times between various warehouse and final delivery point locations. In exercise 7.3, you will calculate warehouse service areas and create a logistics capacity map. Finally, in exercise 7.4, you will simulate the impact and possible response to obstructions or impasses along a road network.

Layer or attribute	Description
FDPs.shp	**Final delivery points**
NAME	FDP town's name
Warehouses.shp	**Warehouse points**
NAME	Warehouse town's name
Status	Status of operations (current or proposed)
Capacity	Warehouse capacity (tonnes)
Regions.shp	**Ethiopia second administrative layer boundaries polygons**
NAME	Region name

Table 7.1 Data dictionary

Exercise 7.1

Building a network dataset

A network dataset is created from the feature classes or shapefiles that participate in the network. A network dataset incorporates an advanced connectivity model that can represent complex scenarios, such as multimodal transportation networks. (Connectivity is a topological property relating to how geographical features are related to one another functionally, spatially, or logically. A multimodal transportation network model is one that incorporates several different types of transportation modes such as road, rail, river, and air.) The network dataset also possesses a rich set of attributes that helps model travel impedance, restrictions, and a hierarchy for the network.

This exercise leads you through the steps for creating and attributing the network dataset that you will use in subsequent exercises. Think carefully about the functions that your network dataset should carry out. Then plan and organize your source data's geometry, topology, and attribute information according to the functions you need performed.

Network routing analysis typically aims to minimize travel cost (or *impedance*); therefore, data sources need to include attribute data that represents travel cost, the cumulative value of which will be minimized over the selected route. You will find that the selection of appropriate types of attribute data is reflected by the design of the UNSDI-T geodatabase schema, which was developed with interurban network analysis in mind.

Create a network dataset

1. **Start ArcCatalog by double-clicking the ArcCatalog icon. Make a connection to the Chapter7 folder within the GISHUM folder. (Use the same Connect to Folder button that you use within the regular ArcMap interface.)**

2. **Enable Network Analyst by clicking Tools > Extensions, and then checking Network Analyst. Click Close.**

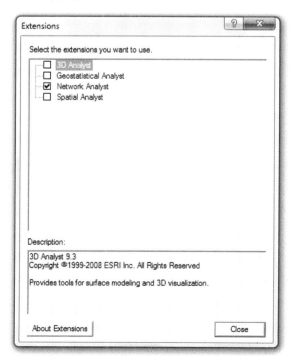

In the Chapter7 folder, you see a personal geodatabase version of UNSDI-T that you studied in chapter 5. Personal geodatabases use Microsoft Access for file storage, and therefore carry the extension *.mdb. It was the original file format for ArcGIS geodatabases, but it is gradually becoming less popular due to file size restrictions and poor scalability. Nonetheless, the UNSDI-T personal geodatabase contains a topologically clean road network of Ethiopia ("Roads" feature class in "Transportation" feature class), attributed according to UNSDI-T standards. The road network includes attributes for road class and road surface, which Network Analyst will use to derive travel impedance parameters.

3. **Navigate to the geodatabase's Transportation dataset, select Roads, and then preview its table.**

Notice that while most of the Roads layer values remain unspecified, the Srf field does contain some specific values. You will base your travel cost model, in this case travel time, on Srf (the road surface type determines travel speed) and Shape_Length.

Your routing analysis will be based on metric distances and speed values, so before you create a network dataset you must project, or modify, the source data by exporting it from its current WGS84 Geographic Coordinate System to a metric projection, such as UTM.

4. **Create a new file geodatabase in the Chapter7 folder, and then name it Ethiopia_Network.gdb.**

You will now export the data from the original personal geodatabase format (*.mdb) to the file geodatabase format (*.gdb). This will give you better storage performance and capability than if you used a personal geodatabase.

5. **Within the new geodatabase, create a feature dataset called Transportation. For projection choose Projected Coordinate System > UTM > WGS 1984 > WGS 1984 UTM Zone 37N.prj.Accept all defaults from there, and click OK.**

6. **Right-click the Transportation feature dataset, and then import the Roads feature class from the original geodatabase, UNSDI-T_ETH_GDB.mdb.**

The WGS 1984 UTM Zone 37N projection is automatically applied to the Roads layer.

7. When your settings appear as follows, click OK.

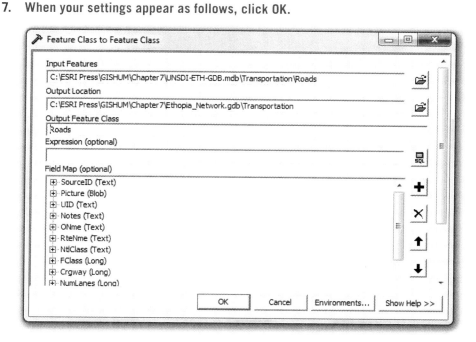

8. Navigate to the new Transportation dataset within Ethopia_Network.gdb and, using the Preview tab, check that the roads were loaded correctly.

9. Right-click within the Transportation feature dataset, and then go to New > Network Dataset.

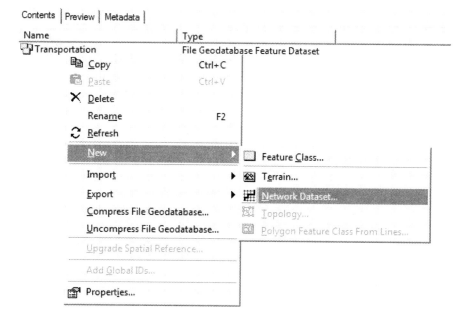

10. Name the network dataset **Roads_ND**, and then click Next.

11. Make sure Roads is checked, and then click Next.

12. Accept the default connectivity settings in the next window, and then click Next.

13. Click Next again since you do not need to modify the connectivity with elevation field data or elevation field settings (as you would if you were modeling road bridge overpasses, for example; these do not occur on Ethiopia's interurban road network).

14. Click No to modeling turns since you are not using them either, then click Next. (Modeling turns would allow you to specify travel behaviors, like expected slowdowns, and restrictions, such as Right Turn Only, at particular junctions.)

Assign impedance values

The next window allows you to specify how Network Analyst processes road feature attributes to derive impedance. You will create one for distance (in kilometers) and one for drive time (in hours).

1. Click Add to create a new network dataset attribute (you will start with the distance attribute). Call it **Distance**.

2. Leave the usage type as Cost since this attribute represents the cost of traveling along the network in terms of distance traveled. Set the units to Meters. Leave the data type as Double, and then click OK.

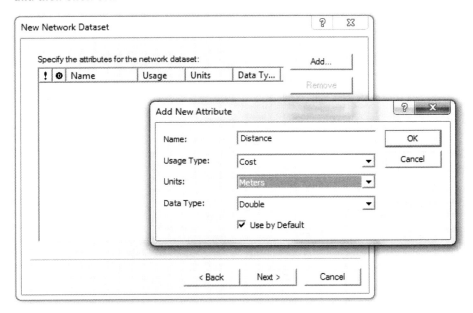

3. Select your new attribute, and then click Evaluators.

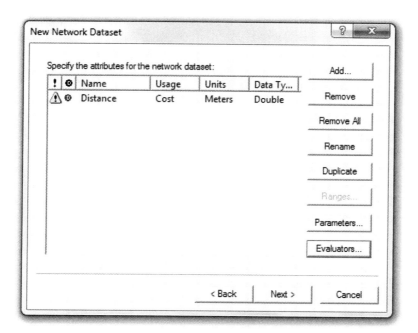

The Evaluators window allows you to specify different attribute values for each feature direction (determined by the line's digitizing). For this exercise, we will assign the same impedance values for the From-To and To-From direction and use only road length and surface type as the determining cost factors.

4. **Under the Type column, select Field for both directions.**

5. **Under the Value column, select Shape_Length for both directions.**

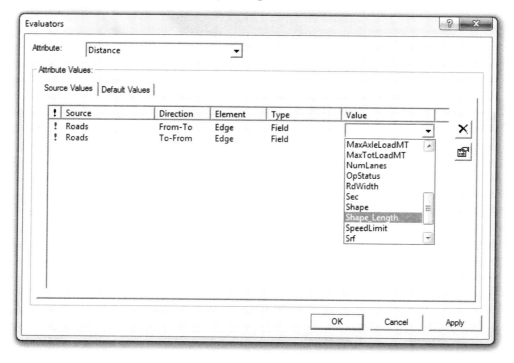

6. **Click OK to return to the Attributes window.**

Next, you will create an attribute for travel time based on the maximum speed possible on different road surfaces.

7. **Click Add. Name the new attribute TTime (a common abbreviation for TravelTime), leave its usage type as Cost, set the units to Hours, leave the data type as Double, and check Use by Default. (The latter specifies that routing solutions will optimize travel time by default, which is usually more appropriate than simply optimizing by distance, especially in developing countries. There, where the shortest route between two points might run along very poor and impractical roads, a detour along higher grade roads is often preferable.)**

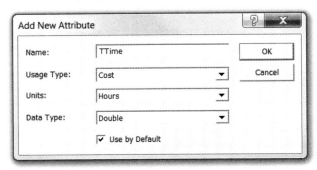

8. **Click OK, then Evaluators.**

9. **Once again, set the Type column to Field for both directions.**

10. **Select one of the directions, and then click the Evaluator Properties 🖻 button.**

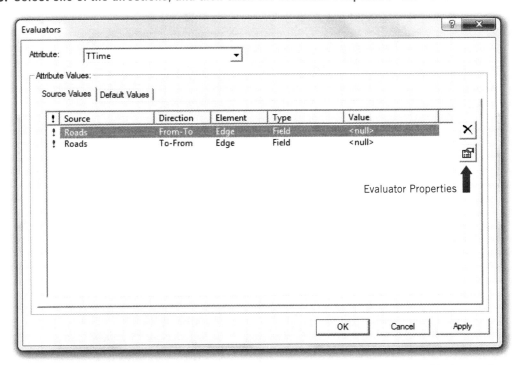

Travel time along an edge (a single road segment) will be based on travel speed (in km/h) along a given road surface type and the edge's distance (time = distance/speed). We will assume the following travel speeds:

Paved roads (Srf = 1): speed = 70km/h
Gravel roads (Srf = 2): speed = 50km/h
Dirt roads and "Unspecified" surfaces(Srf > 2): speed = 20km/h

11. In the Pre-Logic VB Script Code window, type the following code:

```
dim ttime
if [Srf] = 1 then
ttime = [Shape _ Length]/70000
elseif [Srf] = 2 then
ttime = [Shape _ Length]/50000
else
ttime = [Shape _ Length]/20000
end if
```

12. In the Value = text box, type ttime.

13. When your settings appear as follows, click OK.

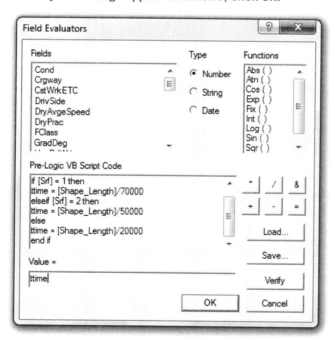

14. Repeat for the other edge direction.

15. When you are done, click OK to close the Evaluators window.

The reason we choose to model travel time this way rather than taking it from a hard-coded value in the attribute table (as we did for Distance, for example) is that road conditions change frequently and travel speeds vary accordingly. By placing this script in the Field Evaluator, changes in road surface are automatically factored into the travel speed calculation every time Network Analyst is run. This becomes especially useful when more attributes are taken into account, such as security and surface condition. It saves us from having to manually recalculate travel time whenever a road condition attribute changes.

16. In the New Network Dataset window, click Next.

17. In the next window, click No (since we will not be establishing driving directions).

18. Click Next.

19. Review the network dataset summary.

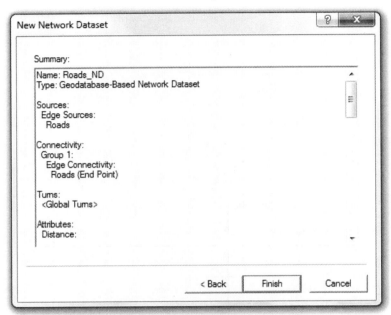

20. Click Finish.

21. Click Yes to build the network dataset (building the network dataset is required anytime you make changes to its source features or parameters).

22. When the new network dataset has been built, close ArcCatalog.

Visualize the network dataset using ArcMap

1. Open a new empty map in ArcMap, navigate to Ethopia_Network.gdb, and add Roads_ND. Click No to displaying all feature classes that participate in Roads_ND.

2. Right-click Roads_ND, and then scroll to layer properties. Click the Display tab.

3. Check Show MapTips, then select TTime in the associated drop-down menu. Click OK.

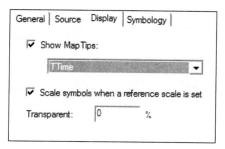

4. Now if you hover the cursor over a given line segment, a MapTip gives you the calculated travel time over that segment. You can also activate the Identify 🛈 button on the main toolbar to view distance and time calculations for any road segment.

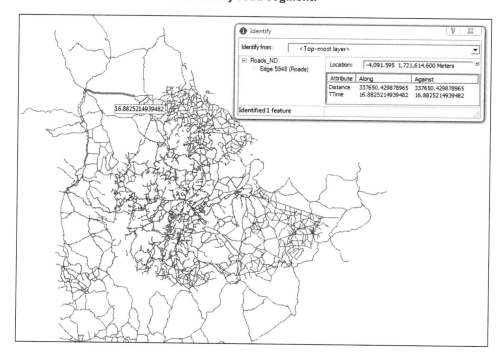

5. You have now created the network dataset on which the rest of this chapter's exercises are based. Close your ArcMap session (you do not need to save the project file at this stage).

Exercise 7.2

Generating an origin-destination matrix

Humanitarian logisticians regard travel time and distance between two locations as fundamental information. Origin-destination (OD) matrices, such as those found in traditional road atlases, are often generated manually in humanitarian operations by selecting a limited number of significant locations, then laboriously researching and entering distances or travel times for each pair. ArcGIS Network Analyst allows such OD matrices to be generated for a much larger set of locations in a fraction of the time required by manual processing.

As mentioned in exercise 7.1, rather than simply showing the shortest route, the matrix can show travel costs along the quickest route, a prohibitively time-consuming calculation when carried out manually. Also, changes to the road network (due to road closures or the arrival of new, more accurate information) simply require a rerun of Network Analyst rather than a recalculation of each individual OD pair. Moreover, several different types of travel costs (distance, time, fuel consumption) can be generated simultaneously, giving logisticians a range of useful criteria on which to base their operational decisions.

Geographic data quality and reliability should be considered carefully when deciding between automated and manual analysis, however. The road network data for many developing countries is appropriate only for "high-pass" analyses. The initial estimate of travel costs provided by such filtered data should be complemented and adjusted with local knowledge.

This exercise will lead you through the steps required to generate an OD matrix between some of WFP's main warehouses and its distribution points (referred to as "final delivery points" or FDPs). Based on the roads network dataset you created in exercise 7.1, you will do the following:

- Learn how to display Network Analyst tools and windows in ArcMap
- Create a new OD matrix group layer and load the data points to be used in the analysis
- Set the OD matrix analysis settings
- Run the Network Analyst solver and visualize the results

Create a new group layer

1. **Start a new ArcMap session.**

2. **Enable the Network Analyst extension (on the main menu, go to Tools > Extensions > Network Analyst). (This needs to be done in ArcMap even if you already enabled Network Analyst in ArcCatalog.)**

3. **Display the Network Analyst toolbar (on the main menu, go to View > Toolbars > Network Analyst). Dock the toolbar somewhere convenient.**

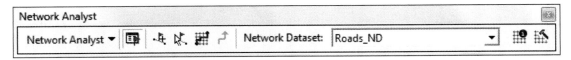

4. **Click the Show/Hide Network Analyst Window** ▣ **button.**

A new frame should appear. You may want to dock it next to the table of contents if it is not already positioned there.

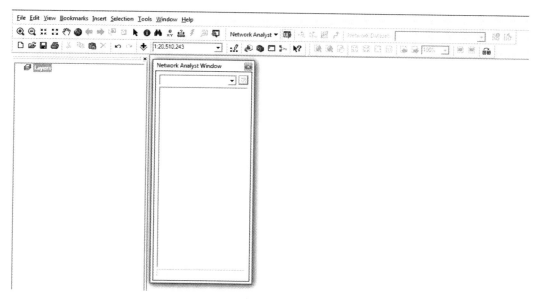

5. **Add the network dataset (Roads_ND) you created in exercise 7.1 to the data frame. This time, click Yes to adding the feature classes that participate in the network.**

6. **From the Chapter7 folder, add the Warehouses, FDPs, and Regions shapefiles.**

7. **Clear the Rds_ND_Junctions layer check box so that Warehouses and FDPs appear more clearly.**

8. **Right-click the Regions layer, and then select Zoom to Layers.**

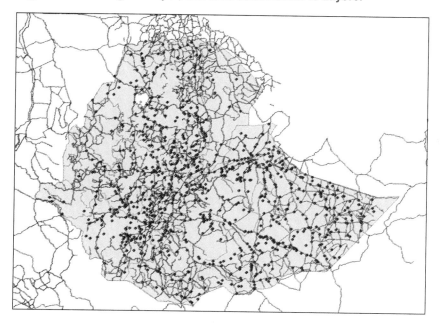

9. **Click the Network Analyst drop-down menu, and then click New OD Cost Matrix.**

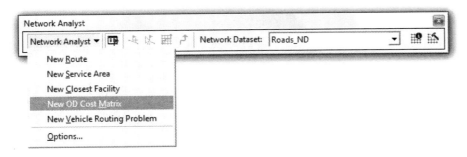

A new group layer called OD Cost Matrix appears in the table of contents, containing an Origins, a Destinations, a Barriers, and a Lines layer. Before running the analysis, you will need to assign features to the Origins and Destinations layers.

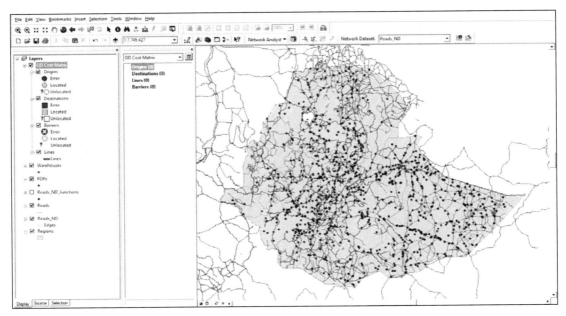

Load origin and destination points

1. Right-click Origins in the Network Analyst frame, then click Load Locations.

2. Select Warehouses from the Load From drop-down menu, making sure the Name Property is set to NAME (this ensures that the Origin points are named according to the warehouse names). Leaving the search tolerance to the default value of 5000 meters, click OK.

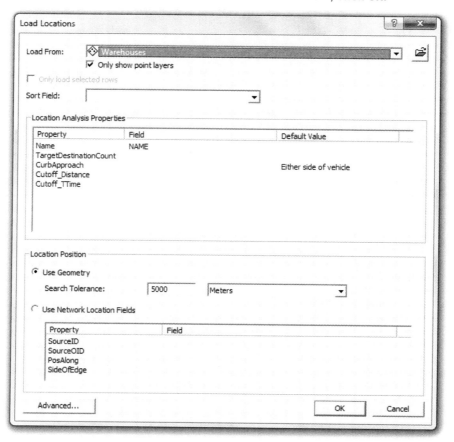

Notice that all warehouses were successfully located on the network (you can see this insofar as they are all symbolized as Located). This means that all warehouses were within the 5,000-meter search tolerance of a network edge feature, and therefore considered connected to the road network.

3. **Follow the same procedure to load destination locations using the FDPs shapefile.**

Once the loading has completed, notice that 219 destination points were not located. You can verify this by opening the Destinations attribute table and selecting all records where Status equals Not Located. These points will be excluded from the OD cost matrix, since they were beyond the 5,000-meter search tolerance.

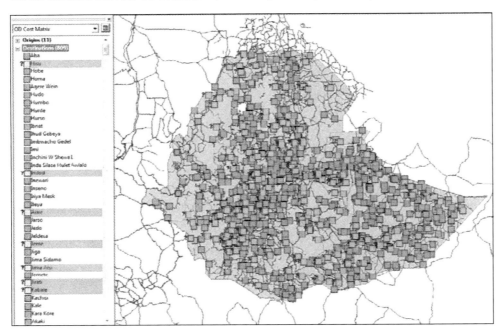

You could force Network Analyst to locate all destination points by increasing the search tolerance if it was absolutely necessary to service every FDP. But for now, keep the 5,000-meter search tolerance and note that a significant number of destinations will be excluded from the OD cost matrix.

Calculate matrix values based upon selected impedance

1. **Click the OD Cost Matrix Properties** 🔳 **button in the Network Analyst frame. This allows you to set a few final parameters before running the analysis.**

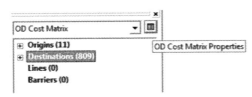

2. **In the Analysis Settings tab, be sure that the impedance is set to TTime (Hours). You chose this as the default in exercise 7.1 to ensure that Network Analyst will use total travel time to determine the shortest route between Origin-Destination (in this case, Warehouse–Final Delivery Point) pairs.**

3. Leave the Default Cutoff Value, Destinations To Find, and Allow U-Turns in their default settings (None, All, and Everywhere, respectively). Set Output Shape Type to None; given the number of destinations you are computing, generating straight lines between all OD pairs would create an unwieldy number of lines.

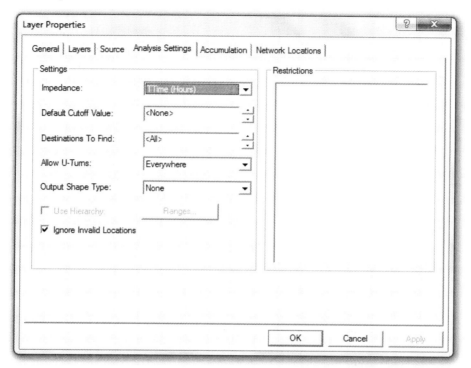

4. In the Accumulation tab, check Distance to ensure that the results display the total distance along the quickest route between OD pairs. As you are using TTime as impedance, it will automatically be included in the results, so there is no need to check its box.

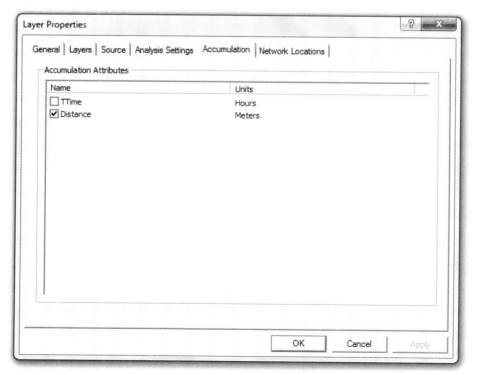

5. **Click OK.**

6. **The OD Matrix analysis is now ready to run, so click the Solve  button on the Network Analyst toolbar.**

7. **If prompted, close the warning box notifying you of the destinations not located.**

8. **Open and inspect the Lines attribute table.**

Notice that the records are grouped by Origin ID, then sorted in ascending order according to the closest destination (see the DestinationRank field). This means that for each origin, you can quickly identify its closest destination points.

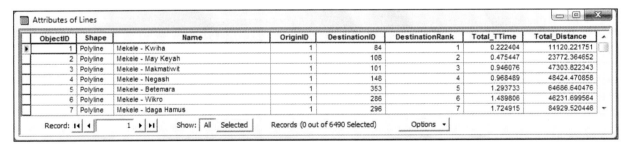

ObjectID	Shape	Name	OriginID	DestinationID	DestinationRank	Total_TTime	Total_Distance
1	Polyline	Mekele - Kwiha	1	84	1	0.222404	11120.221751
2	Polyline	Mekele - May Keyah	1	108	2	0.475447	23772.364652
3	Polyline	Mekele - Makmatiwit	1	101	3	0.946076	47303.822343
4	Polyline	Mekele - Negash	1	148	4	0.968489	48424.470858
5	Polyline	Mekele - Betemara	1	353	5	1.293733	64686.640476
6	Polyline	Mekele - Wikro	1	286	6	1.489806	46231.699564
7	Polyline	Mekele - Idaga Hamus	1	296	7	1.724915	84929.520446

At this stage you have a number of options for packaging the results. You will want to join the Origins and Destinations layers to the Lines layer in order to be able to identify the pairs by name rather than by ID. The OriginID and DestinationID fields of the Lines layer can be joined to the ObjectID field in the Origins and Destinations layers, respectively.

9. Click the Options drop-down button in the Lines attribute table, and then select Joins and Relates > Join.

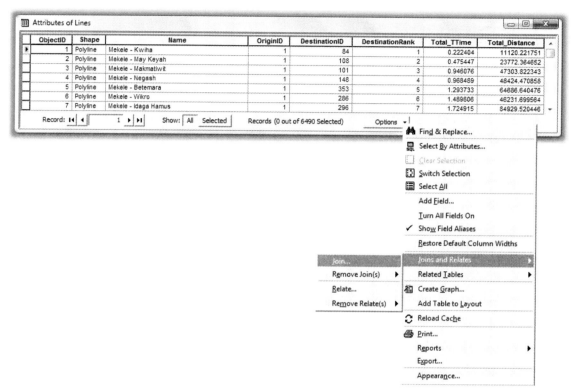

10. Replicate the following Join wizards to associate names with the OriginID, and DestinationID fields with the Lines attribute table.

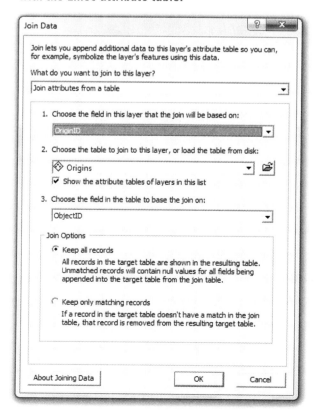

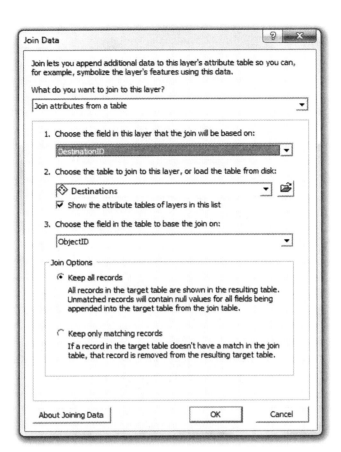

11. Confirm that both joins were successful, then save your work as **GISHUM_C7E2.mxd.**

Produce final matrix using Microsoft Excel pivot tables

You could now create a report by clicking Options in the Lines attribute table, then Reports, then Create Report. However, because these groupings by origin name show all destinations and their respective travel distance and travel time, the report may be several hundred pages in length due to the redundancy in the Destinations field.

A more efficient option is to export the table to a standard spreadsheet package and use pivot tables to create an (m x n) matrix, where m is the number of origin points, and n is the number of destinations. A useful presentation of the results, for example, splits each destination column into distance and travel-time columns.

1. Right-click the Lines layer, and then open the attribute table.

2. Click Options on the bottom right-hand corner of the attribute table.

3. From the Options menu, choose export, and then export the table as a database file named **OD_Matrix.dbf** (when asked if you want to add the new table to the current map, click No).

4. Open the OD_Matrix.dbf file in Microsoft Excel.

5. Quite a few fields appear, more than are necessary for our final pivot table. Delete all of the fields except the four that follow, and rename them as suggested here:

Old Field Name	New Field Name
Total_TTim	Travel Time (Hrs)
Total_Dist	Distance
Name_1	Origin
Name_12	Destination

6. Because the original distance field is in meters, add another field and convert distances from meters to kilometers (i.e., divide column D by 1,000, to two decimals). Your edited table should look like this one:

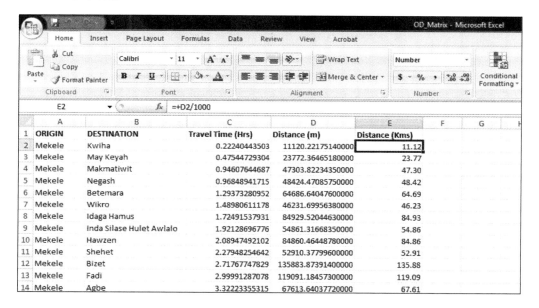

7. To create the pivot table in Excel 2007, click the Insert tab on the main menu of Excel, and in the Tables group, click PivotTable; then click PivotTable again. The Create PivotTable dialog box opens. (If you are using Excel 2003, you will follow a similar process by clicking the Data tab on the main menu, and then launching the PivotTable wizard.)

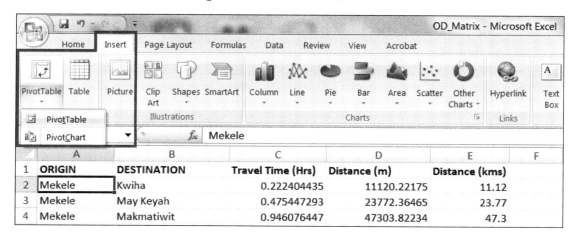

8. Select the range and destination of the pivot table. Use the Range Selection 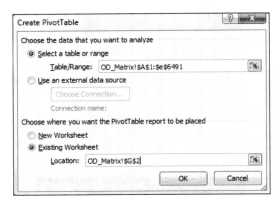 button to ensure that all the tabular values are selected and that the pivot table is placed in an empty column next to the Distance column.

9. Next, select the type and layout of your pivot table field list as shown below. (If you are using Excel 2003, the number of columns is too limited to hold all of the Destinations, so flip the column and row labels that are indicated below.)

10. Finally, close the PivotTable Field List wizard to view your OD matrix. Your table should look as follows:

Row Labels	Column Labels ▾ Aadan-Maanyare	Aba	Aba Bor	Abdela Illubabor	Abera	Abiyu	Abomsa (Tinsae Birhan)	Abosa	Adabo
Addis Abeba									
Sum of Distance (Kms)	1008.006947	535.8478541	471.8043809	442.996674	562.6539804	205.6931755	200.9208867	151.7978641	684.3510622
Sum of Travel Time (Hrs)	24.44677505	8.586684107	7.389564679	6.910201859	11.22399977	5.136362853	3.219227132	2.168540916	15.51342331
Asosa									
Sum of Distance (Kms)	1659.788084	626.7322565	466.0640156	437.2563087	727.5499791	857.4743128	852.702024	803.5790014	1336.132199
Sum of Travel Time (Hrs)	35.68138663	12.99752784	9.130414459	8.65105164	14.50700046	16.37097444	14.45383872	13.40315251	26.7480349
Bahir Dar									
Sum of Distance (Kms)	1478.342763	675.6902656	515.0220247	486.2143178	776.5079882	383.6159973	671.2567022	622.1336797	1154.686878
Sum of Travel Time (Hrs)	32.56086662	13.96084525	10.09373187	9.614369052	15.47031787	9.338406975	11.33331871	10.28263249	23.62751489
Dire Dawa									
Sum of Distance (Kms)	541.7912503	890.9258671	969.8069505	940.9992435	1066.154801	703.6957451	394.0804518	506.8758772	456.234914
Sum of Travel Time (Hrs)	17.7865508	13.65922715	14.5038871	14.02452428	17.96725973	12.25068527	5.978649491	7.24108396	20.17460885
Gambela									
Sum of Distance (Kms)	1689.56222	470.5236715	238.8378473	243.907102	343.4811757	887.2484486	882.4761598	833.3531373	1214.793442
Sum of Travel Time (Hrs)	35.49591783	9.092404613	4.21075433	4.215348105	6.869623514	16.18550563	14.26836991	13.2176837	25.98269629
Gode									
Sum of Distance (Kms)	472.7414197	1512.368799	1591.249882	1562.442175	1687.597732	1325.138676	1015.523383	1128.318809	245.4790683
Sum of Travel Time (Hrs)	16.79100698	26.27675285	27.1214128	26.64204998	30.58478543	24.86821097	18.59617519	19.85860966	12.27395341
Kemba									
Sum of Distance (Kms)	1381.963824	290.9435808	571.2046758	542.3969689	466.1725144	722.5746092	574.8777639	370.4320725	705.8150346
Sum of Travel Time (Hrs)	32.26196703	7.546254228	12.8466573	12.36729449	11.85428681	14.99333421	11.03441911	7.764837629	18.29300237
Kembolcha									
Sum of Distance (Kms)	1197.727713	907.295328	843.2518548	814.4441479	934.1014544	407.6070594	536.1865508	523.2453381	1055.798536
Sum of Travel Time (Hrs)	27.63207404	13.89307659	12.69595716	12.21659434	16.53039226	8.87174596	8.483738957	7.474933401	20.8198158

11. You can adjust the table structure by manipulating the drop-down menus on the row labels. The following version of the OD matrix was generated by removing the Sum of Travel Time (Hrs) data, thereby simplifying the table.

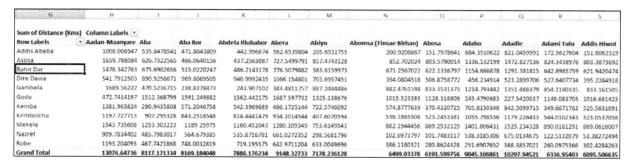

Sum of Distance (Kms) Row Labels	Column Labels ▾ Aadan-Maanyare	Aba	Aba Bor	Abdela Illubabor	Abera	Abiyu	Abomsa (Tinsae Birhan)	Abosa	Adabo	Adadle	Adami Tulu	Addis Hiwot
Addis Abeba	1008.006947	535.8478541	471.8043809	442.996674	562.6539804	205.6931755	200.9208867	151.7978641	684.3510622	821.0459991	172.5627604	151.6062319
Asosa	1659.788084	626.7322565	466.0640156	437.2563087	727.5499791	857.4743128	852.702024	803.5790014	1336.132199	1472.827136	824.3438976	803.3873692
Bahir Dar	1478.342763	675.6902656	515.0220247	486.2143178	776.5079882	383.6159973	671.2567022	622.1336797	1154.686878	1291.381815	642.8985759	621.9420474
Dire Dawa	541.7912503	890.9258671	969.8069505	940.9992435	1066.154801	703.6957451	394.0804518	506.8758772	456.234914	523.2899706	527.6407734	395.2384918
Gambela	1689.56222	470.5236715	238.8378473	243.907102	343.4811757	887.2484486	882.4761598	833.3531373	1214.793442	1351.488379	854.1180335	833.161505
Gode	472.7414197	1512.368799	1591.249882	1562.442175	1687.597732	1325.138676	1015.523383	1128.318809	245.4790683	227.5420037	1149.083705	1016.681423
Kemba	1381.963824	290.9435808	571.2046758	542.3969689	466.1725144	722.5746092	574.8777639	370.4320725	705.8150346	842.5099715	349.6671762	525.5631091
Kembolcha	1197.727713	907.295328	843.2518548	814.4441479	934.1014544	407.6070594	536.1865508	523.2453381	1055.798536	1179.226433	544.0102343	523.0537058
Mekele	1543.735608	1253.303223	1189.25975	1160.452043	1280.109349	753.6149543	882.1944456	869.2532329	1401.806431	1525.234328	890.0181291	869.0616007
Nazret	909.7834402	485.7983017	564.679385	535.8716781	661.0272352	298.5681796	102.6973797	101.7483117	538.3185306	675.0134675	122.5132079	53.38272494
Robe	1193.204093	467.7421868	748.0032819	719.195575	642.9711204	633.0049696	386.1180321	280.8624328	251.6907652	388.3857021	260.0975366	302.4284263
Grand Total	13076.64736	8117.171334	8169.184048	7886.176234	9148.32733	7178.236128	6499.03378	6191.599756	9045.106861	10297.94521	6336.95403	6095.506635

12. Save your finished Excel pivot table.

You have now created an OD matrix for Ethiopia's main warehouses and the final delivery points (FDPs). Such origin-destination matrices have proven to be very useful to logisticians and other field operators for a wide range of applications. The following exercises show how ArcGIS Network Analyst can be used to derive more information not immediately apparent from simple visualization.

What to turn in

If you are working in a classroom setting with an instructor, submit an Excel file containing your OD matrix and any associated analysis.

Exercise 7.3

Analyzing service areas and closest facilities

The service area of a given facility is the polygon enclosing all points more efficiently served by it than by any other facility. As well as knowing the travel cost between a warehouse and a final delivery point, a logistician frequently needs to decide which set of FDPs are most efficiently served by each warehouse. The objective is to minimize transportation costs over an operational area and to anticipate throughput of each warehouse based on the total amount of aid being distributed by its associated set of FDPs. It is only possible to know whether your warehouses are adequate if you know the cumulative needs of the beneficiaries they service.

In this exercise, you will determine the service areas of the warehouses in the Warehouses shapefile and derive their respective throughput requirements. You will do the following:

- Learn to use two new Network Analyst tools: Service Area and Closest Facility
- Learn to join the Network Analyst output layers to source layers in order to derive new information

Create a new service area analysis layer

1. **If not already opened, open GISHUM_C7E2.mxd, and then clear the OD-Cost matrix group layer check box.**

2. **From the Network Analyst drop-down menu, select New Service Area.**

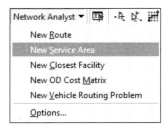

3. **Click Selection on the main menu, and then click Select by Attributes to select all warehouses where Status equals Current.**

The warehouse shapefile contains the locations of warehouses that are currently used in WFP operations, as well as proposed warehouse locations. You will analyze proposed warehouse locations at the end of this chapter, but the remainder of the exercises are based on the current warehouse configuration.

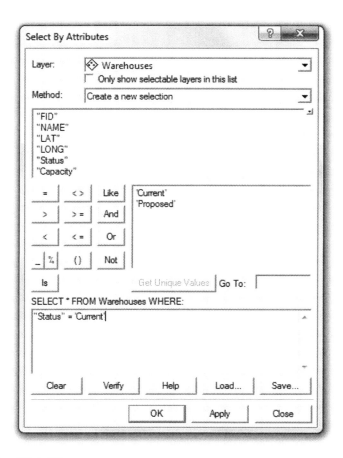

4. Click OK.

5. In the Network Analyst window, right-click Facilities, and then select Load Locations.

6. Select Warehouses in the Load From drop-down menu. Be sure that the "Only load selected rows" check box is checked, make sure the Name Property in the Location Analysis Properties grid is set to Name, and then click OK.

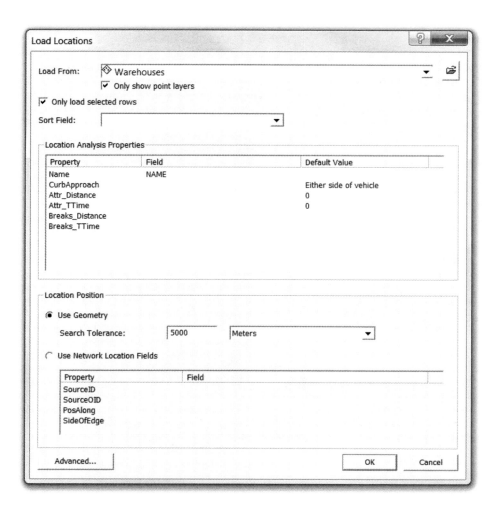

7. Clear all selected features (on the main menu, go to Selection > Clear Selected Features).

Generate service polygons based upon selected impedance

You will now create four service areas indicating the range of final delivery points that can be most efficiently served by the five warehouses operated by the World Food Programme in Ethiopia.

1. Click the Service Area Properties ▦ button in the Network Analyst window.

2. Click the Analysis Settings tab, and then set the Default Breaks to 1, 5, 10, and 15; this will generate a series of doughnut polygons around each warehouse representing the areas that can be reached in 1, 5, 10, and 15 hours, respectively.

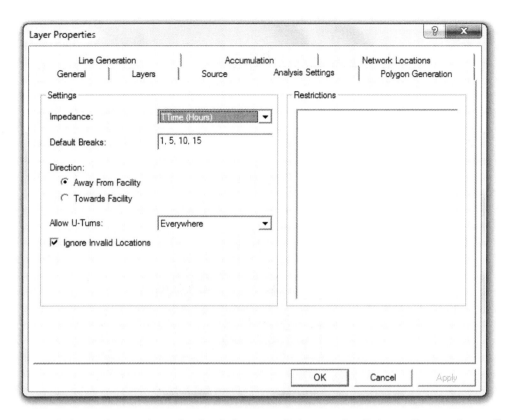

3. In the Polygon Generation tab, check Generate Polygons. Set Polygon Type to Generalized; uncheck Trim Polygons. Set Multiple Facilities Options to Not Overlapping. Set Overlap Type to Rings.

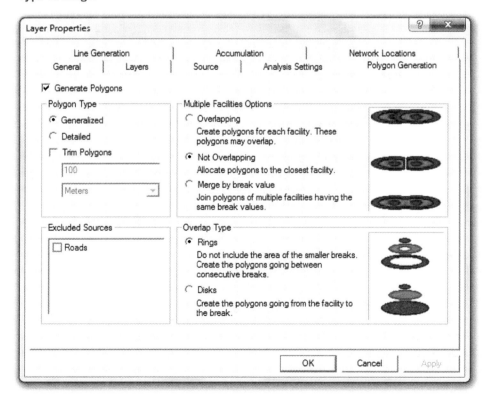

4. **Click OK, then click the Solve button.**

A quick inspection of the resulting polygons, which represent break values of 1, 5, 10, and 15 hours of travel time, show that the majority of FDPs are located beyond a five-hour drive from a warehouse.

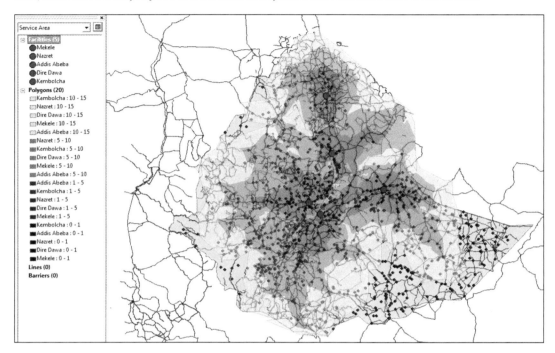

You can quantify this by spatially joining the FDP layer with the polygon file, then computing statistical summaries on the ToBreak field of the join output; or by spatially selecting FDPs falling within specific break polygons. Alternatively, symbolizing the service area polygons by Facility ID makes apparent which warehouses serve the most FDPs. Using this latter option, you can quantify not only the number of FDPs served by each warehouse but also its total aid requirement. You can carry out this procedure more efficiently, however, using the Closest Facility tool of Network Analyst, which is what you will do next.

5. **Save your work as GISHUM_C7E3.mxd.**

Create a new closest facility analysis layer

1. **Clear the Service Area group layer check box.**

2. **From the Network Analyst drop-down, choose New Closest Facility.**

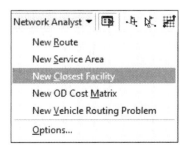

This tool is designed to allocate sets of incident points to specific facilities based on minimum cost paths. (Here, "cost" refers to whichever impedance attribute the user has defined: distance, time, monetary transport cost, etc.). For example, in emergency services (its most common application), the New Closest Facility tool allows a dispatcher to quickly identify which fire stations (the facilities) will be quickest to respond to a particular house fire (the incident). In our case, we will treat FDPs as our incidents and warehouses as our facilities, and identify which FDPs should be served by which warehouse. We will then compute the total throughput requirement of each warehouse based on its allocated FDPs.

3. **Repeat steps 3 to 6 on pages 250 to 251 to load current warehouses into the Facilities of the New Closest Facility group layer. Remember to select only warehouses whose status is current, not proposed.**

4. **In the Network Analyst window, right-click Incidents, and then select Load Locations.**

5. **Now it is time to load the destination points. Select FDPs in the Load From drop-down menu. Ensure that the Name Property in the Location Analysis Properties grid is set to Name.**

When you carried out this procedure to derive an OD matrix (exercise 7.2), you were more concerned with the geometric accuracy of your results. This time you want to ensure that the service load of each warehouse is accurately reflected. We are more interested in computing the complete set of FDP aid requirements associated with a given warehouse than in the actual distance between any FDP and the warehouse. By failing to account for its service load, you risk underestimating the total throughput capacity requirement of a warehouse, which can lead to logistical bottlenecks. Therefore, you must take into account all FDPs by increasing the search tolerance used by Network Analyst. Then, if some road or trail unknown to our database does link faraway FDPs to the main road network, even if the procedure comes up with inaccurate absolute travel cost values, it will be allocating FDPs to warehouses correctly. Increasing the search tolerance simply creates a little contingency in case some routes are not reflected in our network dataset.

Optimize delivery points based upon selected impedance

1. A quick analysis indicates that the most remote FDP is slightly less than 29 kilometers from the road network, so set the search tolerance to 29000 meters. Click OK.

2. Open Closest Facility Properties.

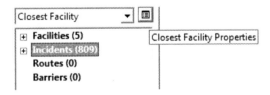

3. In the Analysis Settings tab, be sure that Impedance is set to TTime, Default Cutoff Value is set to None, and Facilities To Find is set to 1. (This way, you will be computing shortest drive time, including all FDPs no matter how distant they are from a warehouse, and identifying only one closest warehouse per final delivery point.)

4. In addition, set Output Shape Type to Straight Line. (As well as reducing computing time, this setting produces straight lines between a warehouse and its associated FDPs, giving a much clearer schematic view of each allocation.)

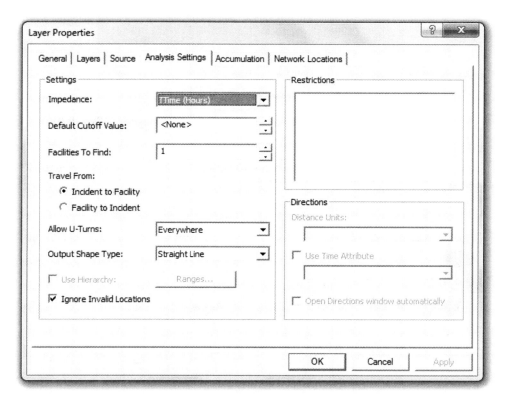

5. Under the Accumulation tab, check Distance, then Click OK to save your settings. Exit the Closest Facility properties window.

6. Click the Solve button.

7. Turn the Service Area group layer back on to verify that both results are consistent with each other. FDPs allocated to a particular warehouse should all fall within that warehouse's service area polygon.

Changing the symbology of the Routes layer (within the Closest Facility group layer) and Polygons layer (within the Service Area group layer) is the most efficient way to determine this.

8. Right-click Routes, and then go to Properties > Symbology. In the Show box, click Categories, then Unique Values. For the Value Field, choose FacilityID. Click Add All Values.

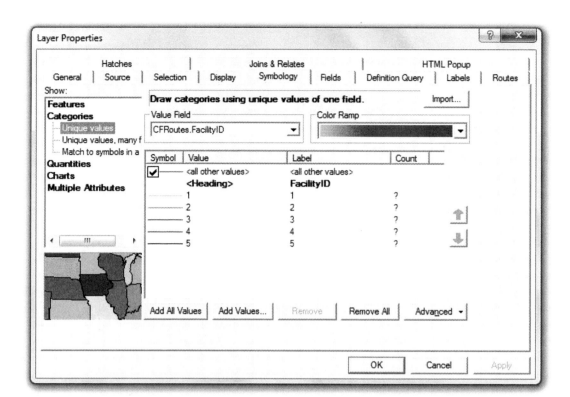

9. Repeat the same symbolization process for the Polygons layer, reduce the symbol size of the Incidents layer, and label the Facilities layer. Your map should be similar to the following example.

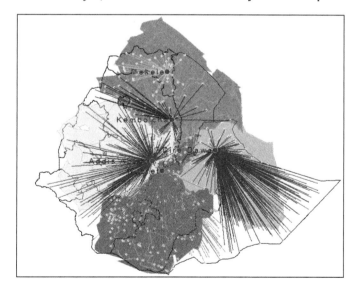

Summarize delivery facility statistics

You will now compute and summarize the total commodity requirement for each warehouse, based on its respective set of FDPs, and check that each warehouse has sufficient throughput capacity to cope with this demand.

To do this, you need to join the FDP attribute table (containing a monthly metric ton requirements attribute, Reqs) to the Warehouses attribute table (containing a throughput capacity attribute, Capacity). Since there is no direct connection between these two layers, you will use the Routes, Facilities, and Incidents layers to provide the intermediary linkage, as illustrated in the following image.

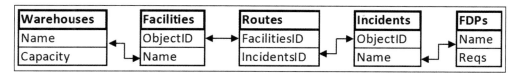

Each route in the Routes layer references its origin facility (Warehouse) by the ObjectID of the corresponding Facilities point. You will therefore use the Routes layer's FacilityID and the Facilities layer's ObjectID to create the first join in the chain.

To avoid confusion when making the joins indicated, you may want to give the Name field a unique alias in the Warehouse, Facilities, Incidents, and FDPs layers. That will make it easier to render each of your joins correctly. (Open each layer's properties and go to the Field tab to change field aliases, as illustrated on the Facilities layer in the following image. The alias has been changed from Name to FacilityName.)

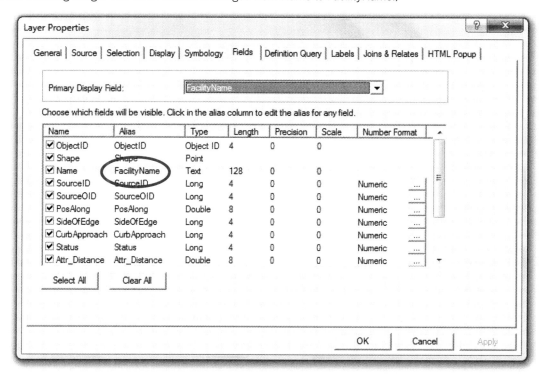

1. **Right-click Routes, and then select Joins and Relates > Join. Use the same approach to join Incidents to Routes (using the Routes layer's IncidentID and the Incidents layer's ObjectID). Set the parameters as shown in the following images.**

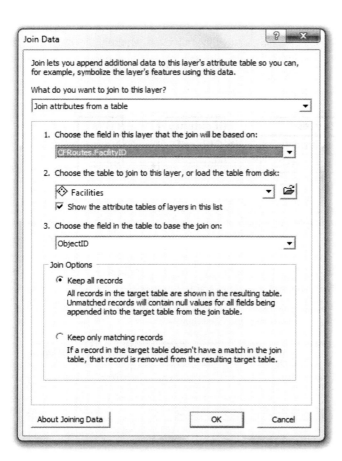

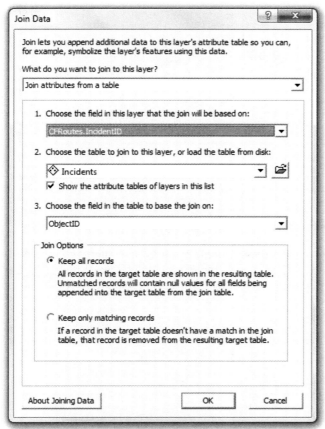

2. Now join the FDPs layer and the Warehouses layer to the Routes table. Use the Name attribute of the Facilities and Incidents layers to link to the Name attribute of Warehouses and FDPs layers, respectively.

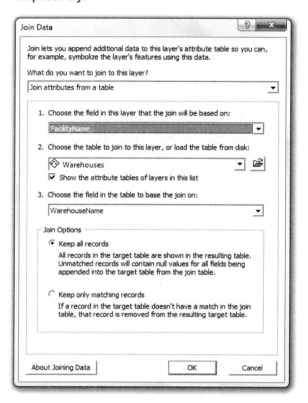

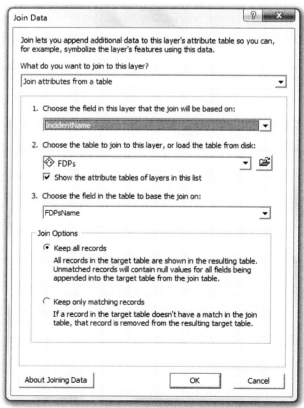

3. Confirm that your joins are successful by opening the Routes attribute table. To reveal the original layer and name of each field, turn off the field aliases by clicking the Options drop-down button and clearing the Show Field Alias option in your attribute table. This will label each column in the following format: {layer's name:field's name}

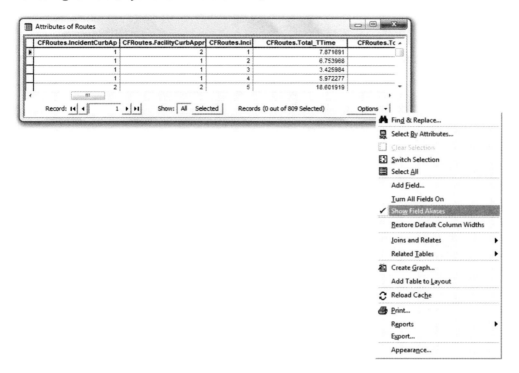

Your Routes attribute table should contain all the attributes necessary to compute each warehouse's total throughput requirement and compare it to its capacity. If so, you have successfully joined four tables to the Routes layer. If not, remove and re-create your joins to ensure that the Routes table is properly connected with the Warehouses and FDPs layers.

4. Right-click the field heading WarehouseName, and then select Summarize. (You may have to scroll about three-quarters of the way across the table to see the WarehouseName field.)

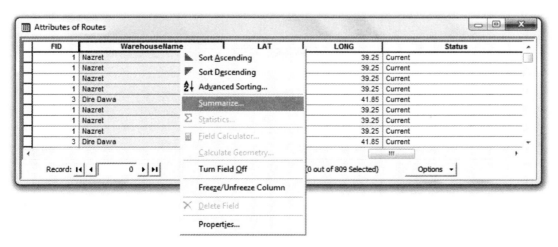

5. Under box 2 of the Summarize wizard, scroll down to FDPs.Reqs, click the expand (+) button, and check Sum. This will add the total FDP requirements associated with each warehouse.

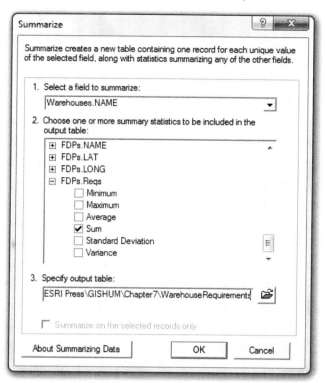

6. Save the output table as **WarehouseRequirements.dbf** in the Chapter7 folder. Click OK.

7. When prompted, click Yes to add the table to the map project, and then open it. You can now see the total commodity requirements for each warehouse, as well as the total number of FDPs served by each warehouse.

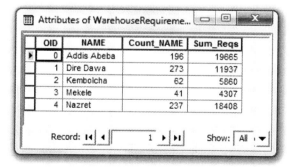

Create a logistics capacity map

You will now symbolize each warehouse with column charts showing throughput requirements against capacity.

1. Start by joining the WarehouseRequirements.dbf table to Warehouses, using Name as the join attribute.

2. Open the Warehouses properties and symbolize the layer with a bar chart using the Capacity and Sum_Reqs attributes.

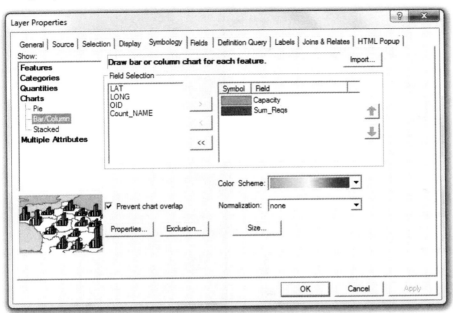

3. You may need to drag the Warehouses layer to the top of the table of contents in order to clearly see the graphs. A quick glance at the map shows that all warehouses are operating within capacity, except for the one in Addis Ababa. (Note that the graphic shows the official spelling of the Ethiopian Mapping Authority, not the conventional English spelling.)

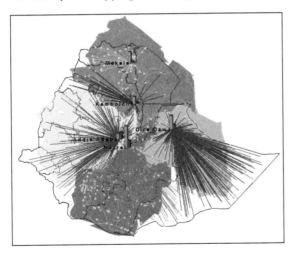

4. Save your work as **GISHUM_C7E3.mxd**.

What to turn in

If you are working in a classroom setting with an instructor, submit an electronic or printed copy of your Exercise7.3.mxd map display.

In this exercise you learned how to combine Network Analyst outputs with source features to derive information useful in strategic logistical planning. In the next exercise, you will use Network Analyst's Closest Facility tool to derive contingency plans and what-if scenarios.

Exercise 7.4

Creating a contingency plan

Whether because of political or military conflict, natural disasters, or predictable seasonal effects, the logistics infrastructure and transportation network underlying a humanitarian operation is constantly subject to disruption: Warehouses may be looted, communication lines cut off, or ports and airports damaged. Whole sections of roads may be closed down due to landslides, insecurity, heavy rains, or countless other events. The exact timing of a contingency is rarely predictable, but that something will happen is always a case of when, not if. What you can do in advance is assess what its impact on logistical operations might be.

GIS can be used to simulate the impact of such scenarios. In this exercise, you will carry out a comparative analysis of population accessibility between a current-state scenario and a simulated road closure. A contingency plan like this can help prepare for the effects of disruptions on the transportation network and estimate resulting additional costs, delivery times, or storage requirements. In this exercise you will do the following:

- Learn to derive operational statistics from closest facility outputs
- Learn to place barriers along the network to simulate blockages

Derive operational statistics

1. Open **GISHUM_C7E3.mxd** (if it is not already open), and then remove all joins to the Routes layer.

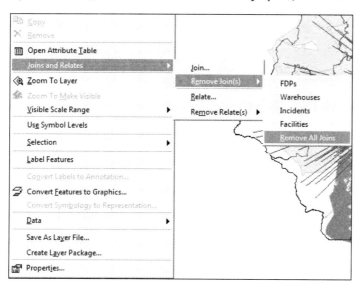

2. Join the Routes layer to the Incidents layer using their IncidentID and ObjectID fields, respectively. Click No, if prompted to create an index.

3. Symbolize Incidents by FacilityID using the same color scheme you applied to Routes in the previous exercise. You may wish to turn on the roads layer to provide context, and turn off the service area polygons to improve legibility.

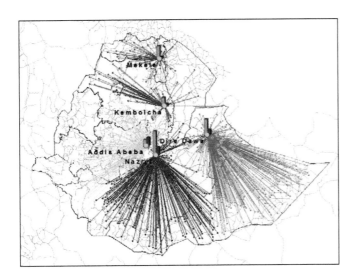

4. Select the Somali region from the Regions layer.

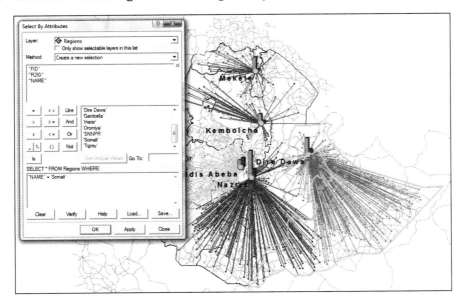

5. Select Incidents that are completely within the selected region, Somali.

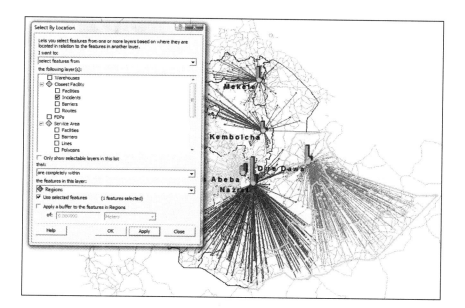

6. Open the Incidents table, and then click the Show: Selected button.

7. Turn off field aliases, right-click the field heading CFRoutes.Total_TTime, and click Statistics.

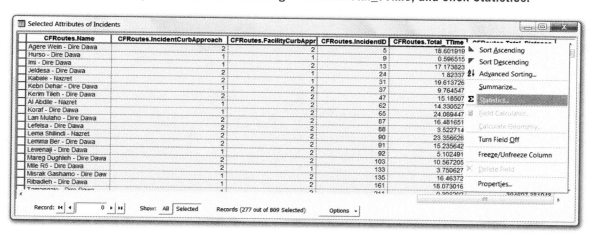

8. Make note of the Sum and Mean travel time (in hours) to the FDPs in the Somali Region:
   1. Sum: _____
   2. Mean: _____

9. Once you have taken note of these figures, close the statistics window and minimize (don't close) the attribute table.

Add hypothetical route barriers

You will now add two hypothetical road blockages in the area of Dire Dawa (the easternmost warehouse).

1. Zoom in on the area around Dire Dawa, and then clear the Routes layer check box to make the area more clearly visible.

2. In the Network Analyst window, select the Closest Facility network, and then highlight Barriers(0).

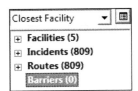

3. In the Network Analyst toolbar, select the Create Network Location Tool ⊹ button.

You will now create hypothetical road blocks, which will force Network Analyst to bypass using alternative routing strategies (detours).

4. Place barriers at the two locations shown in the image.

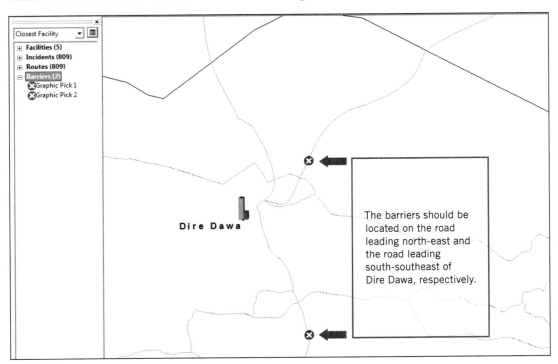

5. Turn the Routes layer back on, and then label the Incidents by Name.

Reoptimize delivery points

1. Click the Solve button. Observe the change in FDP-Warehouse allocation. Also notice that one FDP (Baraaq) is now unreachable (close the Network Analyst Messages window once you have read it).

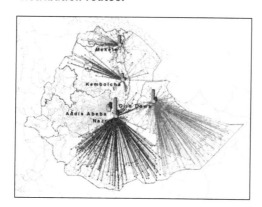

2. Restore or reopen the Incidents attribute table (it should still show only FDPs within the Somali region). Run the statistics again on CFRoute.Total_TTime. Note the new Sum and Mean travel times to the FDPs:
    1. Sum: _____
    2. Mean: _____

3. The road closures have caused a 26 percent increase in total drive time (presumably, with a commensurate increase in travel costs), as well as a 26.5 percent increase in average travel time to individual FDPs. You can see that the barriers exert a substantial impact on optimal distribution routes.

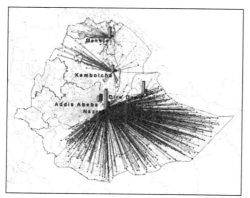

This chapter introduced a fairly sophisticated use of ArcGIS—network analysis—and you are quickly developing into a skilled spatial analyst.

Assignment 1

Based on the procedures covered in exercise 7.3, derive the new throughput requirements for each warehouse, and then check that surrounding warehouses are not now exceeding capacity.

What to turn in

If you are working in a classroom setting with an instructor, submit the following:

1. Your OD matrix pivot table in Microsoft Excel format from exercise 7.2.
2. Your map display at the end of exercise 7.3.
3. Your calculation of the increased total and average travel time from exercise 7.4.
4. Your results from the assignment.

Assignment 2

You have just received some good news: there is additional funding to operate three more WFP warehouses. Determine the best locations, with the goal of minimizing the total travel time for relief distribution in Ethiopia. Then determine the best locations, changing your goal to minimizing total transportation costs. Prepare a short report for the WFP's country manager that describes the trade-off between travel time and travel cost for the three additional warehouse locations.

Note that your warehouse feature layer includes a number of potential warehouses (Status = Proposed) distributed across the country. Your challenge, therefore, is to discover what configuration of warehouses provides the greatest reduction in travel time.

As an additional dimension to your analysis, consider estimating total transport costs, as follows:

Transport Cost to each FDP (in Ethiopian Birr) = 36.31 + (0.333 x Distance(Km) x Total Requirements(MT))

What to turn in

If you are working in a classroom setting with an instructor, submit an electronic copy or printout of your report.

OBJECTIVES

Identify potential locations for humanitarian services
Calculate slopes from topographic (SRTM) terrain data
Transform multiple input layers into a single, "qualified" output layer
Automate repetitive analytical processes using ArcGIS ModelBuilder

Chapter 8

Selecting sites with multiple spatial criteria

The large-scale displacement of people is an unfortunate reality in humanitarian emergencies. Such displacement can last many months or years before affected populations are able to return to their homes. Some are never able to return, and temporary camps become semipermanent settlements for successive generations of displaced people.

Deciding where to deliver humanitarian assistance can be a complex spatial challenge —making it an ideal application of GIS. Learning how to exploit GIS to short-list potential locations, based upon a set of predefined criteria, is a core skill for the humanitarian GIS specialist. In this chapter, you will apply spatial analysis to select suitable locations for an internally displaced persons (IDP) camp in Uganda, a country that has a large population of displaced people.

A forgotten crisis: The Ugandan Civil War

In 2006, the UN Office for Coordination of Humanitarian Affairs (OCHA) reported widely shared concerns about the neglected humanitarian crisis in northern Uganda. At that time, approximately 1.6 million people had been displaced by a conflict waged by the Lord's Resistance Army, a rebellious movement that was especially brutal in its abduction and militarization of children. The Ugandan Civil War is often categorized as a "forgotten humanitarian crisis"; it has received relatively little attention despite the fact that the number of internally displaced persons (IDPs) in that region is comparable to that in Darfur, Sudan.

According to OCHA, conditions in the IDP camps across northern Uganda were poor, despite improvements over the previous year and a half. The quality of medical care, water, and sanitation were far below standards. To complicate matters further, the civil war created an insecure environment for humanitarian organizations attempting to provide aid in the region. (See "Uganda: Child soldiers at centre of mounting humanitarian crisis" from *10 Stories the World Should Hear More About*, Department of Public Information, United Nations 2006, available at http://www.un.org/events/tenstories/06/story.asp?storyID=100.)

Before proceeding, familiarize yourself with the mass human displacement caused by the Lord's Resistance Army conflict. See the Web resources listed at the end of this chapter.

Scenario: Determining suitable camp sites for internally displaced persons

Now turn your attention to the Ugandan situation as if you were a GIS specialist engaged in humanitarian assistance. Based upon your knowledge of good practices in shelter selection, as well as directives from the field and headquarters, you will search for locations in Uganda that match the following suitability criteria:

Gradient: Shelter sites require good drainage. If the site's gradient is too flat, the area will be prone to flooding during rainy seasons. With standing water acting as a breeding ground for vector-borne diseases such as malaria, poor drainage can lead to a variety of health risks. It can allow human and animal waste to contaminate inhabited areas and water and food supplies. On the other hand, if the gradient is too steep, the difficulty and cost of operating a large shelter can become prohibitive, partly due to the risks imposed by slope failure. Therefore, you decide to consider only *areas with a gradient of between 2 and 7 degrees.*

Road accessibility: Aid agencies require year-round access in order to ensure ongoing camp management, including essential goods and services, such as emergency rations, water, and building and medical supplies. Setting up a temporary or permanent site at a location that is only seasonally accessible can exacerbate an already difficult situation. Therefore, you decide to consider only *areas within 5 kilometers of a major road.*

Market accessibility: Often, the reality is that displaced populations stay in shelters for a much longer duration than they (or their host communities) expected at the time of displacement. They need to be able travel out to seek better livelihoods and commercial markets. Such accessibility should be a consideration in site selection. Therefore, you decide only to consider *areas within 25 kilometers of a major settlement.*

Safety/security: Shelters should be located away from local hazards such as land mines, unexploded ordnance, flood channels, and urban threats that might jeopardize IDP security. When hazards are unavoidable, the emphasis is placed upon reducing the vulnerability of inhabitants (barriers, safety zones, security patrols, evacuation routes, etc.). Yet the first choice is always to minimize risk by avoiding hazards. Therefore, you decide to limit your consideration to *areas farther than 10 kilometers from a known UXO site and farther than 5 kilometers from a major settlement.*

These criteria represent only a few factors to consider when establishing new humanitarian infrastructure. You can use many other criteria to optimize the location of a camp, depending upon the specific conditions and constraints of an emergency situation.

Solving spatial problems: A conceptual model

Obviously, you approach any complex spatial problem with a conceptual model that ensures your analysis will provide a reliable and successful solution to that problem. In this chapter, you will follow a conceptual model for conducting suitability analysis of IDP camps in northern Uganda, which uses a series of steps.

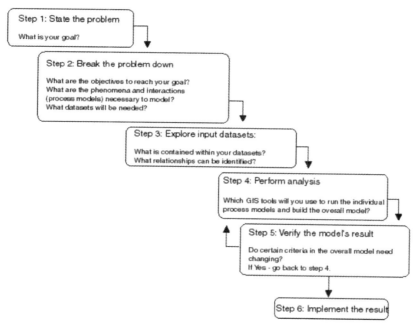

You can use GIS tools to analyze the suitability of IDP camps, according to the thinking outlined in the following six steps of the conceptual model.

Step 1: State the problem

To solve a spatial problem, first state the problem you are trying to solve and the goal you are trying to achieve. Then visualize the type of results you want to produce.

In this chapter, your challenge is to find the best locations for new IDP camps in northern Uganda. Ultimately, you want a map showing potential sites (ranked best to worst) that could be suitable for building new camps. This is typically called a "ranked suitability map" because it shows a relative range of values indicating how suitable each location is on the map, according to the criteria used for the analysis.

Step 2: Break the problem down

Once the problem is stated, break it down into smaller pieces until you know what steps are required to solve it.

Define your analytical objectives by considering how to measure the suitability of IDP camps, taking into account humanitarian standards and conventional best practices. In this chapter, you want to identify all areas with the following characteristics:

- Enough slope to ensure adequate drainage of seasonal rains
- Proximity to major roads, so that humanitarian agencies can more easily deliver assistance
- Access to commercial markets, so that camp inhabitants can seek better livelihoods through trade
- Safe from Explosive Remnants of War (ERW), such as land mines and unexploded ordnance (UXO)
- Outside urban areas, due to land constraints

While these five analytical objectives suffice for your analysis, any criteria helpful in identifying site suitability could be added. Such criteria might include land use and tribal or ethnic relations, as well as proximity to humanitarian services, police stations, or army bases.

Step 3: Explore input datasets

Once you have broken down the problem into a series of objectives and process models, and chosen the datasets you will need, explore these datasets. Understand the content of the input datasets so that you know which attributes within, and between, datasets are important for solving the problem.

The fact that spatial data may be limited during an unfolding humanitarian emergency makes it all the more important to carefully study the data you do have. What about it could be used to perform meaningful analysis? Search for ways to use the data to locate camps, to weight for certain input attributes, and to improve the modeling process.

Step 4: Perform analysis

Once you have decided on your objectives, the elements and their interactions, the process models, and what input datasets are needed, you are ready to perform the analysis.

Two options emerge for performing analysis to find the best locations for new IDP camps:

Option A: Create a suitability map that shows a suitability value for every location in your region of interest. This involves ranking the values of each input data layer according to its suitability for hosting IDP camps. For example, each class of a land-use map could be assigned a value of 1 through 5, with 5 designating land areas most suitable for new IDP camps. You can apply a similar suitability scale to every other input data layer, with a value of NoData applied to mask off areas that should be excluded from analysis. Once reclassified according to this suitability scale, the values of each input layer can be collectively analyzed and weighted according to their relative importance.

Option B: Query your datasets using the above criteria to obtain a Boolean result of true or false. This involves finding every location that totally satisfies all suitability criteria. For example, use the Raster Calculator to find all locations that have a slope gradient of between 2 and 7 degrees and are within 5 kilometers of roads, within 20 kilometers of urban markets, more than 10 kilometers from known UXO hazards, and more than 5 kilometers from urban centers. The result is a Boolean true or false map of locations that either match or do not match that set of suitability criteria.

What distinguishes options A and B is the flexibility factor: When you query your data to get a Boolean true or false map (option B), you exclude areas of partial suitability. Locations must meet all criteria in order to be deemed suitable. If your analysis warrants more flexibility, choose option A; this will produce a raster map where every location (cell) has a suitability value.

Step 5: Verify the model's result

If possible, verify the accuracy of your spatial analysis by visiting the potential sites in the field. You may find that your analysis did not account for something important.

For example, you may discover a toxic waste dump upwind of the site that is producing foul odors, or a new restriction on building camps that wasn't considered in the suitability analysis. For the analysis to identify potential IDP camp locations correctly, this new field-level intelligence needs to be factored in.

Step 6: Implement the result

The final step in the spatial model is to implement the results by preparing guidance documents to decision makers that explain the method of your analysis and rank the suitability of potential sites.

In this case, the action is to find the most suitable locations for IDP camps: Once a conceptual model of IDP camp suitability analysis has been built and tested, you can automate all of the geoprocessing steps using ArcGIS ModelBuilder. This enables the analysis to be quickly repeated whenever updated input data layers become available. Also, using this proven process for site selection allows you and others to replicate the analysis in other areas of interest.

Layer or attribute	Description
Districts.shp	**Uganda third administrative layer boundaries polygons**
DNAME	District name
Pop	District population
Populated_Places.shp	**Uganda settlements points**
NAME	Settlement name
Roads.shp	**Uganda roads lines**
TYPE	Primary, secondary, tarmac
UXOs.shp	**Uganda unexploded ordnance sites points**
DISTRICT	District at risk
SRTM_Uganda	**Uganda Shuttle Radar Topographic Mission grid**

Table 8.1 Data dictionary

Exercise 8.1

Generating slope data using Surface Analysis

Among the most significant limitations in past humanitarian emergencies has been the lack of topographic models for countries in the developing world. That has changed in recent years, as a result of the Shuttle Radar Topography Mission (SRTM), conducted by NASA in 2000. Reliable terrain data is now available, at no cost, for most of the earth's surface with a resolution of 90 meters (3 arc seconds). SRTM marks a tremendous improvement over the previous global model standard (GTOPO30), and in many countries, is quite often better than the nationally available topographic information. (You can read more about the SRTM on its official Web site: http://www2.jpl.nasa.gov/srtm.)

1. **Start ArcMap, and then open GISHUM_C8E1.mxd.**

You should see a raster layer named SRTM_Uganda. You will use this terrain data to create a slope layer from which you will extract areas that are suitable to camp construction.

2. **Enable the Spatial Analyst extension (on the main menu, go to Tools > Extensions > Spatial Analyst).**

3. **Display the Spatial Analyst toolbar (on the main menu, go to View > Toolbars > Spatial Analyst). Dock the toolbar somewhere convenient.**

4. **On the Spatial Analyst toolbar, choose Surface Analysis > Slope from the Spatial Analyst drop-down menu.**

5. **Populate the Slope dialog box as follows:**

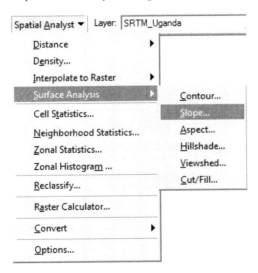

- Input surface: SRTM_Uganda
- Output measurement: Degree
- Z factor: 0.00000898
- Output cell size: Accept default value of 0.000833333
- Output raster: Slope (ESRI Grid format)

6. When your settings appear as follows, click OK.

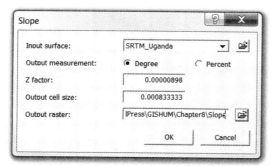

What is the z-factor?

The z-factor is a conversion factor that adjusts the units of measure for the vertical (or elevation) units when they are different from the horizontal coordinate (x,y) units of the input surface. The z-factor is the number of ground x,y units in one surface z-unit. If the vertical units are not corrected to the horizontal units, the results of surface tools will not be correct.

The z-values of the input surface are multiplied by the z-factor when calculating the output surface. If the x-, y-, and z-units are all the same (in feet, for example), the z-factor is 1. This is the default value for the z-factor. For another example, if the vertical z-units are feet and the horizontal x,y units are meters, you would use a z-factor of 0.3048 to convert the z-units from feet to meters (1 foot = 0.3048 meter).

The correct use of the z-factor is particularly important when the input raster is in a spherical coordinate system, such as decimal degrees. It is not uncommon to perceive the output from hillshade as looking peculiar if the input surface raster is not in a projected coordinate system. This is due to the difference in measure between the horizontal ground units and the elevation z-units. Since the length of a degree of longitude changes with latitude, you will need to specify an appropriate z-factor for that latitude.

If the x,y units are decimal degrees and the z-units are meters, the appropriate z-factors for the following latitudes are

Latitude	z-factor
0	0.00000898
10	0.00000912
20	0.00000956
30	0.00001036
40	0.00001171
50	0.00001395
60	0.00001792
70	0.00002619
80	0.00005156

Because Uganda sits on the equator, you used the z-factor for a latitude = 0.

7. Click **OK** when the process is complete. (This may take a few moments to process.)

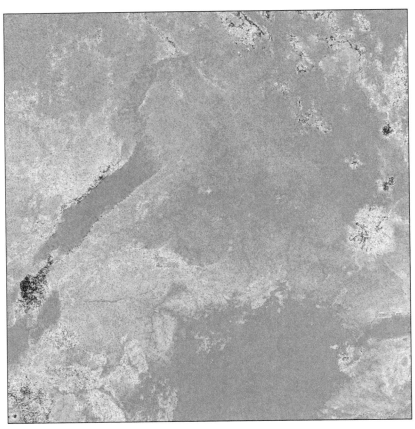

The new slope will appear in your data frame when complete. By default, the slope raster is symbolized, using the natural breaks classification method, into nine categories. Because your organization has decided that the ideal gradient for site selection is between 2 and 7 degrees, you will resymbolize the slope raster into three ranges of slope: less than 2 degrees, between 2 and 7 degrees, and greater than 7 degrees.

8. Right-click the **Slope** layer, and then navigate to **Properties > Symbology.**

9. Change the number of classes to **3**, and then click **Classify.**

10. Change the classification method to **Manual**, click **OK**, then set the classification brackets to the three ranges of slope.

11. When your settings appear as follows, click OK to close the Classification window.

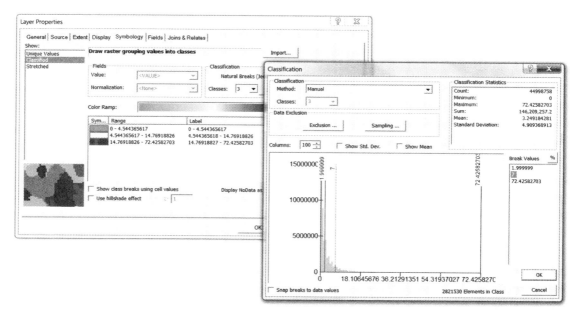

You have created three categories of slope gradient: < 2 degrees, 2–7 degrees (our ideal category), and > 7 degrees.

12. Click OK to close the Layer Properties window. Your raster will look like the following image.

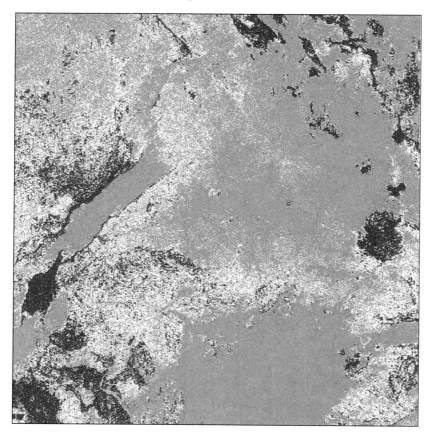

Exercise 8.2

Determining site suitability

Now that you have prepared the slopes dataset, you can begin to prepare the other input data layers necessary to determine site suitability. You will generate these data layers using proximity analysis, and then overlay analysis, in order to isolate only those regions of Uganda that are within 5 kilometers of a road and 25 kilometers of a market, and farther than 10 kilometers from a UXO site and at least 5 kilometers from urban insecurity.

1. **Open GISHUM_CE2.mxd. You should see the correctly symbolized Slope raster, as well as feature layers for roads, populated places, UXOs, and district boundaries throughout Uganda.**

2. **Turn off the Districts and the Slope layers for the moment.**

3. **Click the Toolbox** 📦 **button to access ArcToolbox. Navigate to the Analysis toolbox, then the Proximity toolset, and then the Buffer tool. Launch the Buffer tool wizard.**

4. **Populate the Buffer wizard as follows:**

   • Input features: Roads.
   • Output feature class: C:\ESRIPress\GISHUM\Chapter8\RoadsAccess.shp.
   • Distance: Linear unit of 5 kilometers.
   • Dissolve Type: ALL.
   • Accept all other defaults.

5. **When your settings appear as follows, click OK.**

6. Close the completion window, and then confirm that the roads buffer is consistent with your expectations. Depending on your data settings, the output should resemble the figure below.

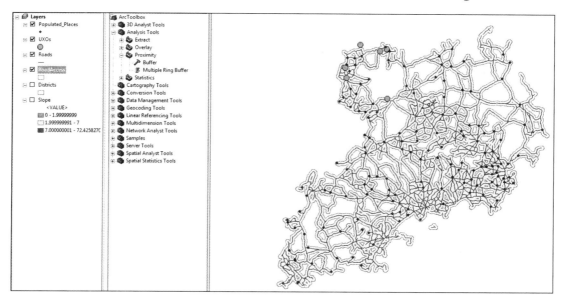

7. Now perform another buffer analysis of the Populated_Places layer. Name the output feature class **MarketAccess.shp**. Set a buffer distance of **25** kilometers, and choose Dissolve Type as All in order to merge the output polygons into a single feature.

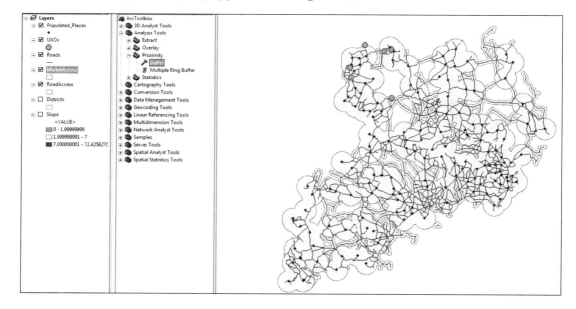

Create an intersect using overlay analysis

Your combined output shows the overlapping RoadsAccess and MarketAccess data buffers. You will now generate a combined overlay, isolating those areas that have both road and market accessibility.

1.   **In the Analysis toolbox, navigate to the Overlay toolset, and then launch the Intersect tool wizard.**

2.   **Add RoadsAccess and MarketAccess to your input features, and name the output feature class RdMktAccess.shp. Accept the remaining default settings, and then click OK.**

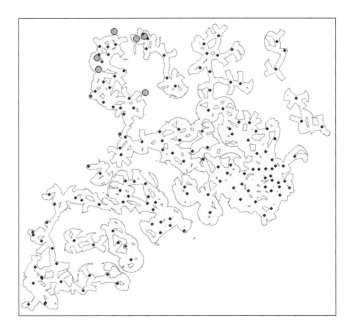

You have now transformed the two original input data layers into a single layer that shows areas offering adequate access to transportation and commercial opportunities. Unfortunately, it also includes areas at risk of UXOs and urban threats, which you will now extract from the RdMktAccess layer.

3. **Perform two more buffer analyses: one of the Populated_Places layer, in which you name the output feature class UrbanHazard.shp and set a buffer distance of 5 kilometers; and another of the UXOs layer, in which you name the output feature class UXOHazard.shp and set a buffer distance of 10 kilometers. In both cases, choose Dissolve Type as All in order to merge the output polygons into a single feature.**

You have now transformed the two original input data layers into layers showing areas considered hazardous. The following figure displays the two hazard layers on top of the RdMktAccess layer.

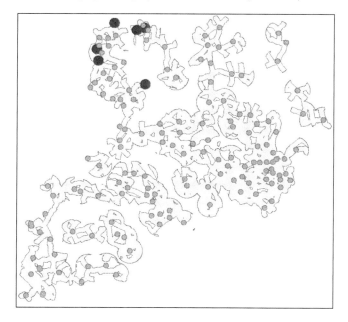

Extract areas using the Clip tool

Your final operation is to erase excluded areas from the RdMktAccess layer; that is, those areas that coincide with the UXO and urban hazard zones. Instead of using the Clip tool provided in the Analysis > Extract toolset, you are going to perform your clip operation as part of an editing session. This alternative approach allows you to discard the clipped features, effectively "erasing" them from the RdMktAccess layer.

1. **Activate the Editor toolbar (from the main menu, select View > Toolbars > Editor). Dock the Editor toolbar conveniently on your desktop.**

2. **From the Editor drop-down menu, choose Start Editing.**

3. **From the Target layer drop-down list, select RdMktAccess.**

4. **Using your mouse, click any UXO hazard symbol in your map display to select UXOHazard as the Clip layer. (If you have not already done so, you may wish to turn off the Populated Places, UXOs, Roads, Districts, and Slope layers to make this step easier.)**

5. Return to the Editor drop-down menu; the Clip option should be active. Select Clip.

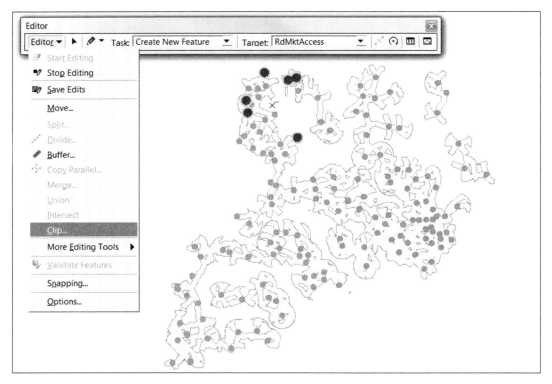

6. In the Clip wizard, leave the buffer distance as 0 (remember this clip layer is already a buffer zone).

7. Choose the option "Discard the area that intersects." (We are using the Editor toolbar version of the clipping function because it gives us this option of erasing the UXO hazard areas from the RdMktAccess layer.)

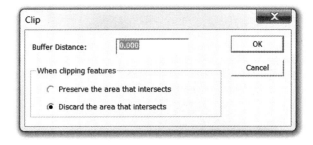

8. Click OK. Once the clip operation has finished, turn off the UXOHazard layer to confirm that the RdMktAccess layer has been edited correctly.

9. Repeat the clip operation, using the UrbanHazard layer. You will need to activate the Edit tool on the Editor toolbar before selecting an urban hazard symbol in your map display.

Your resultant RdMktAccess layer should resemble the following figure.

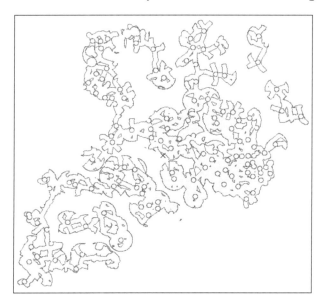

10. **Click Stop Editing on the Editor drop-down menu, and then select Yes to save your edits.**

Your final step in determining suitable IDP camp sites is to isolate those areas that fall within the RdMktAccess layer and have a slope of between 2 and 7 degrees. Though this can be done in several ways; here you will use the Raster Calculator to generate a Boolean test of the criteria (option B in the chapter's introduction).

First, convert your RdMktAccess feature class into a raster layer.

11. **Open ArcToolbox, and then navigate to the Conversions toolbox. Launch the Feature to Raster tool wizard in the To Raster toolset.**

12. **Select RdMktsAccess as the input feature, accept the default Field value, and name the output raster AccessRaster. Select the Slope layer under Output cell size to ensure that the AccessRaster uses the same cell size. (You may manually set the Output Cell Size to "90" in order to maintain the 90-meter resolution of SRTM data.)**

13. **When your settings appear as follows, click OK.**

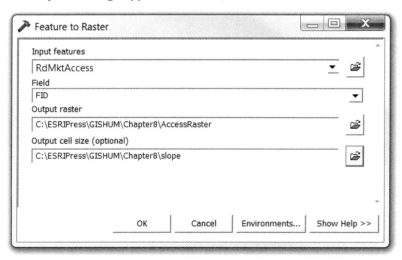

14. Close the operation status window once the feature-to-raster conversion is complete. Then turn off all layers except for the new AccessRaster layer.

You are now ready to perform the Boolean test using your two raster layers.

15. On the Spatial Analyst toolbar, select Raster Calculator from the drop-down menu.

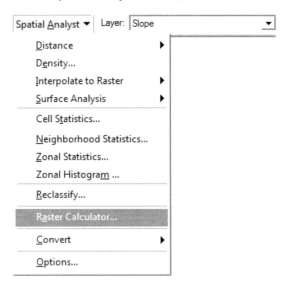

16. You should see your two layers listed in the Raster Calculator wizard. Build the following Boolean expression to compute those areas falling within the AccessRaster and having a slope of between 2 and 7 degrees: **AccessRaster = 0 And Slope >= 2 And Slope <= 7**.

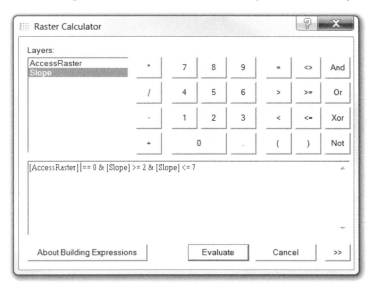

17. Click Evaluate. You now have a raster layer with two Boolean field values: 0 (False) and 1 (True).

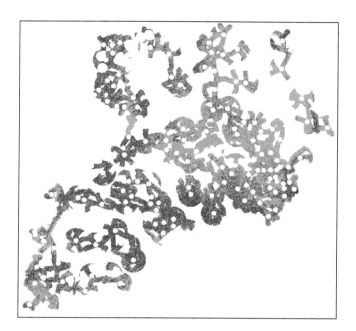

In order to use the results for future analysis, you will now convert the raster calculation into a polygon shapefile, and then compute the area served by each qualifying polygon. This allows you to exclude areas too small to host the anticipated camp population.

18. **Open ArcToolbox, and then navigate to the Conversions toolbox. Launch the Raster to Polygon tool wizard in the From Raster toolset.**

19. **Select Calculation as the input feature and the default field of VALUE. Name the output raster SuitabilityTest. Clear the Simplify polygons option to preserve the input raster cell edges.**

20. **When your settings appear as follows, click OK.**

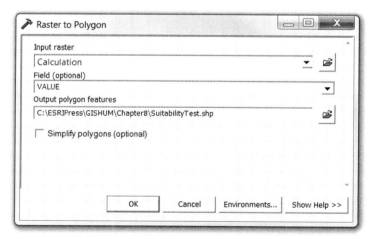

You now have a feature layer containing polygons of suitable (GRIDCODE=1) and unsuitable (GRIDCODE=0) locations for IDP camps.

21. **Open the attribute table to inspect the results. Do they seem reasonable?**

22. From the main menu, click Selection > Select By Attributes in order to identify polygons that are suitable for camp operations.

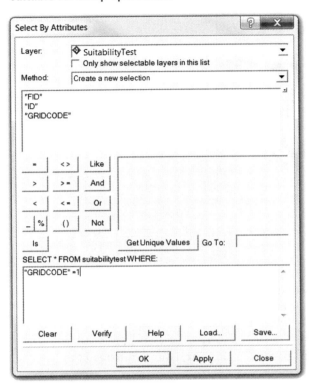

23. Turn on the Districts layer.

24. Right-click the SuitabilityTest layer, and then click Data > Export Data. You are now ready to save the selected features as a separate layer. Use the same coordinate system as the data frame, and then name the output layer **SuitableAreas**.

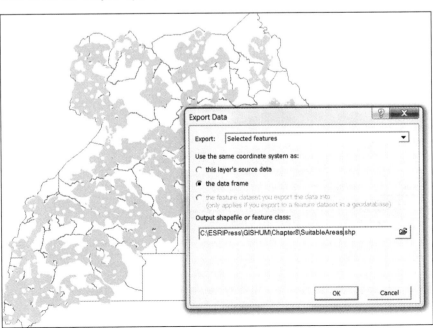

25. Click OK, and click Yes when asked if you want to add the exported data to your map.

Before concluding the analysis, find out the size of each potential camp location so you can approximate the number of displaced people that could be hosted at each site.

26. Open the new layer's attribute table. Delete the GRIDCODE field, since it is no longer informative.

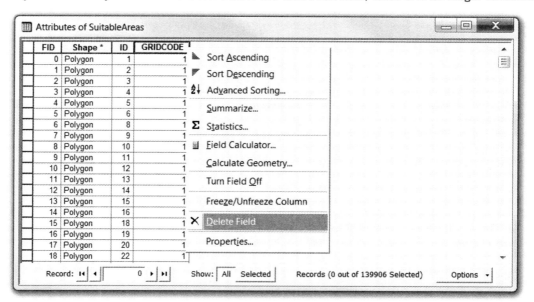

27. Now, using the Options drop-down menu, add a new Field called Area. Select Long Integer as the field type, and give the field a precision of 20. This limits the field value to no more than 20 digits, including decimal places.

28. Click OK.

29. To calculate the area of each suitable site, right-click the new field title Area, and then select Calculate Geometry. Click Yes if asked whether you want to calculate outside of an edit session.

30. Choose to calculate Area in Square Meters, using the default coordinate system.

31. When your settings appear as follows, click OK.

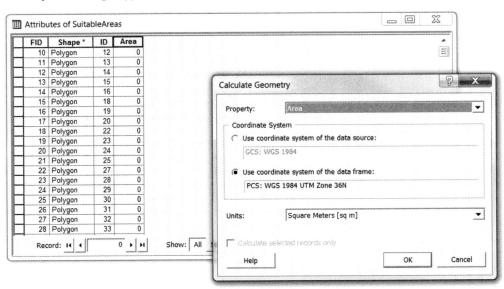

Finally, estimate the number of people that can be accommodated at each site, based upon a benchmark of 45 square meters per person (the recommended standard for shelter planning).

32. Add another field to the SuitableAreas attribute table.

33. Name the new field Capacity, choose Long Integer as the Type from the drop-down menu, and set a Precision value of 20.

34. To calculate the carrying capacity for suitable sites within this district, right-click the field name Capacity, and then click the Field Calculator button.

35. Type the required expression to calculate Capacity = Area/45. When your settings appear as follows, click OK.

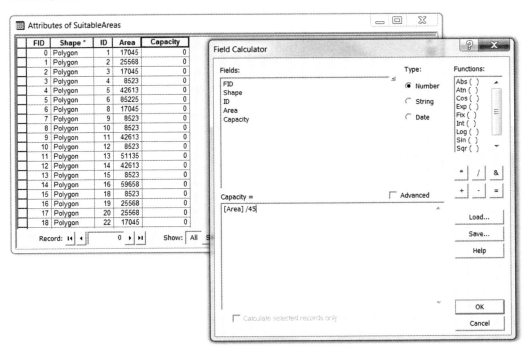

Examine the capacity values for each suitable campsite area.

FID	Shape *	ID	Area	Capacity
0	Polygon	1	17045	379
1	Polygon	2	25568	568
2	Polygon	3	17045	379
3	Polygon	4	8523	189
4	Polygon	5	42613	947
5	Polygon	6	85225	1894
6	Polygon	8	17045	379
7	Polygon	9	8523	189
8	Polygon	10	8523	189
9	Polygon	11	42613	947
10	Polygon	12	8523	189
11	Polygon	13	51135	1136
12	Polygon	14	42613	947
13	Polygon	15	8523	189
14	Polygon	16	59658	1326
15	Polygon	18	8523	189
16	Polygon	19	25568	568
17	Polygon	20	25568	568
18	Polygon	22	17045	379
19	Polygon	23	25568	568
20	Polygon	24	8523	189
21	Polygon	25	110794	2462
22	Polygon	27	42613	947
23	Polygon	28	8523	189
24	Polygon	29	8523	189
25	Polygon	30	17045	379
26	Polygon	31	17045	379

Record: ◄◄ ◄ 0 ► ►◄   Show: | All Selected | Records (0 out of 139906 Selected)   Options ▾

The capacity values suggest that many polygons in the SuitableAreas layer are probably too small to be seriously capable of hosting a camp for thousands of IDPs. Of course, this crude initial analysis should now be verified through field visits and inspection of satellite imagery and applicable documents in order to confirm which sites are most suitable for camp operations.

Your turn

Create a map of all the sites that may be suitable for between 10,000 and 20,000 people within the district of Gulu. (**Hint:** To choose only those polygons with a calculated capacity to support that range of IDPs, you need to clip the SuitableAreas layer using the Districts layer. Then use Select By Attributes to choose only those polygons with a calculated capacity to support 10,000 to 20,000 IDPs.)

Once you are satisfied with your results, generate a table of site locations and capacity. To learn about ArcMap's ability to produce reports automatically, from the main menu, go to Tools > Reports > Create Report.

What to turn in

If you are working in a classroom setting with an instructor, submit an electronic copy or printout of your suitability analysis.

Exercise 8.3

Automating site selection using ModelBuilder

The geoprocessing tools in ArcMap are often used in sequence to perform spatial analysis. Sometimes many steps are involved, making it difficult to keep track of tools used, datasets involved, and parameters defined within the overall workflow.

ModelBuilder is an application within ArcMap that automates and documents your geoprocessing workflows. Within ModelBuilder, you construct model diagrams from the data and geoprocessing tools needed for your analysis or workflow. Once you build the model, you can run it once, or save it and run it again using different input data.

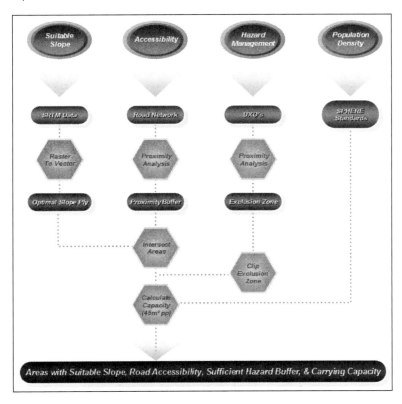

ModelBuilder can be used to automate simple tasks that you frequently perform or to construct very complex analytical processes involving several data sources and geoprocessing tools. Models are an excellent way to create and save visual, sharable, and reusable GIS workflows.

At the highest level, models contain only three things: elements, connectors, and text labels.

1. *Elements* are the data and tools you work with.
2. *Connectors* are the lines that connect data to tools.
3. *Text* labels can be associated with the entire model, individual elements, or individual connectors.

Elements carry their own labels (different from text labels) that you create and associate with models, elements, and connectors. A tool element, for example, is labeled with the name of the tool, but you can change this label.

A tool plus its data is called a *process*. Processes can be in different states: ready to run, has been run, and not ready to run.

Model parameters are data elements that you want to appear on the tool dialog box so that a user can enter values for the data elements. This site selection process will be replicated many times in different parts of the world. ModelBuilder is an application ideally suited to repeatedly perform these geometric processes on datasets for any situation. Once criteria for site selection have been determined and a sequence of steps within ArcMap has been validated to identify suitable locations, you can incorporate these steps into a model for repeated use.

In this section of the exercise, use ModelBuilder to create a model to replicate the geoprocessing workflow you performed in the previous exercises.

Create a new Site Suitability toolbox

1.   From the Chapter8 folder, open GISHUM_C8E3.mxd.

2.   Open ArcToolbox.

3.   Right-click any white space inside ArcToolbox, and then choose New Toolbox. Name the new toolbox **Site Suitability Tools**.

4.   Right-click the Site Suitability toolbox, and then choose New > Model. This should automatically open a new model window.

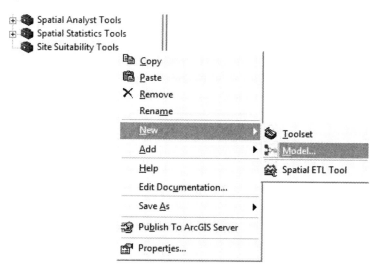

5. Position the model window so that you can see the ArcMap table of contents and the ArcToolbox contents simultaneously. (You may have to resize the widths of both content tables to do this successfully.)

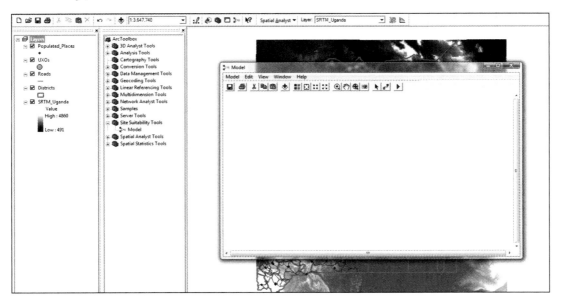

Build the model with analytical tools in ArcToolbox

Not all of the geoprocessing operations performed earlier in the chapter can be automated by ModelBuilder. For example, because the ArcView and ArcEditor licenses do not include the ArcGIS Erase tool, we used the Edit Session Commands from the Editor toolbar to discard hazardous areas from our RdMktAccess layer. Unfortunately, operations performed within an editing session cannot be incorporated into ModelBuilder. Therefore, we will build a model that automates the data preparation needed in order to conduct that operation manually.

1. In the table of contents, click the SRTM_Uganda data layer, and then drag it into the model window.

2. From ArcToolbox, expand the Spatial Analyst toolbox and the Surface toolset. Select the Slope tool, and then drag it into the model window.

3. On the ModelBuilder toolbar, click the Add Connection 🔧 tool.

4. Note that the cursor now looks like a little magic wand. Click the SRTM_Uganda data element, and then drag the cursor to the Slope tool element. You have just created a process!

5. If you double-click the Slope element, the slope wizard will open, allowing you to confirm the input and output rasters, output measurement units, and the desired z-factor. Save the output layer to the ModelBuilder folder in your Chapter8 dataset, and then type a z-factor of **0.00000898**, which is the value recommended for equatorial latitude conversions (see exercise 8.1).

6. When your settings appear as follows, click OK.

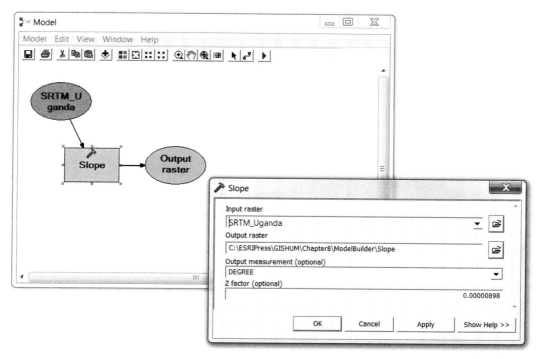

You have now added a process to your model that will prepare a slope raster using an input SRTM data layer. Next, let's add processes to prepare each of the RoadsAccess, MarketAccess, UXOHazard, and UrbanHazard buffers.

Note: You can organize your model's layout anytime by clicking the Auto-layout  tool.

7. Next, drag the Roads layer from your ArcMap table of contents into the model window.

8. From ArcToolbox, expand the Analysis Tools and the Proximity toolset. From the Proximity toolset, click the Buffer tool, and then drag it into the model window.

9. On the ModelBuilder toolbar, click the Add Connection tool.

10. Click the Roads data element, and then drag the cursor to the Buffer tool element.

11. Double-click the Buffer tool element to launch the Buffer wizard. Use the same buffer properties as before but save the Output Feature Class to the ModelBuilder folder in your Chapter8 dataset.

12. When your settings appear as follows, click OK.

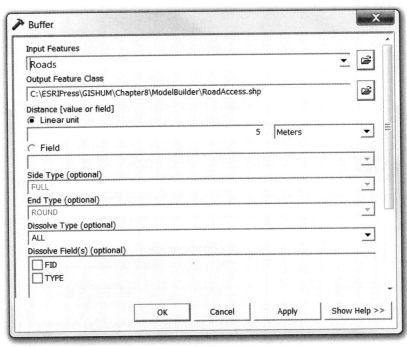

You have now added another process to your model to automatically generate a road buffer called RoadAccess.shp.

13. Repeat steps 1 through 12 on pages 294 through 296, using the respective properties you employed in exercise 8.2, to create the MarketAccess, UXOHazard, and UrbanHazard buffers. Because the Populated_Places layer is used to determine both market accessibility as well as urban hazard, you can connect that data element to both buffer processes.

14. Now add to the model a process that analyzes the intersection of the RoadAccess and MarketAccess layers.

15. Drag the Intersect tool from ArcToolbox > Analysis Tools > Overlay into your model window, and then connect the RoadAccess and MarketAccess data elements as the input layers.

16. Click the Auto-layout button to tidy up the layout. Your model should look similar to the following figure.

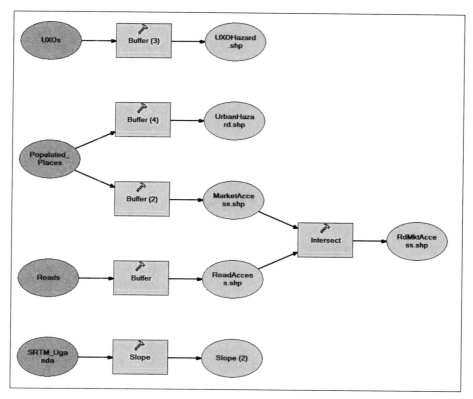

You have now automated most of the data preparation steps needed to test for what areas are suitable for IDP camps.

Run the model

1. Click the Run ▶ button on the ModelBuilder toolbar.

2. As the Model runs through the workflow, the icon of the active process will be highlighted in red.

3. Verify that each of the expected output layers were generated and added to the ModelBuilder folder in your Chapter8 dataset.

You are now ready to perform the manual erase operation using Editor's Clip tool. You could then build another model in your Site Suitability toolbox that automates the feature-to-raster conversion of the RdMkAccess layer, the raster calculation, and then raster-to-polygon conversion that you manually performed in exercise 8.2

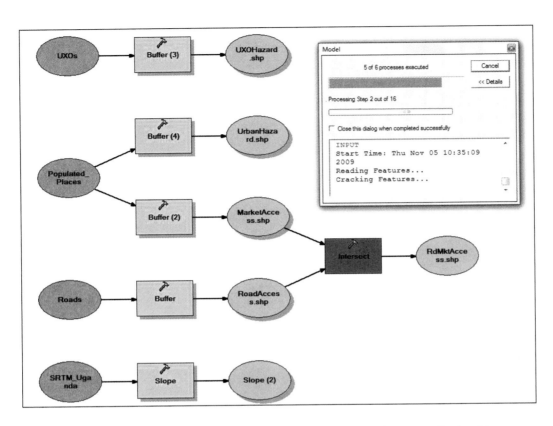

The model can be used to repeat this process for other countries or regions, once the input layers are correctly prepped. The user simply double-clicks each icon in the model, updating file names or paths for the new region. If additional steps (or more advanced criteria) are required, then the model can be edited and saved.

Your turn

Use ModelBuilder to automate another geoprocessing sequence, from this or any previous chapter. Try to automate as many steps as possible!

In this chapter you have acquired several important ArcGIS skills within the context of site selection, but with applicability to other missions as well. Take a moment to list a few of the countless other situations that might require employing ArcGIS to analyze terrain data, perform data transformations, or automate processes using ModelBuilder.

Keep in mind that your first responsibility as a GIS specialist is to ask, *What would be the most helpful types of maps and analysis given the priorities and constraints of my customers and the unique conditions of this emergency?* Providing timely, insightful products is crucial to improving the success of humanitarian operations with GIS.

Assignment

You have developed the core skills for a new GIS assignment: to build a file geodatabase and to develop a geoprocessing model capable of identifying suitable locations for a new 10,000-foot (3,050-meter) runway. Longer runways are sometimes necessary in enabling large aircraft to deliver humanitarian cargo. Your mission is to recommend possible locations to build such a runway in any country of your choice.

Design a set of suitability criteria for airfield development, gather all necessary spatial and nonspatial data, and build a file geodatabase. Then apply your selection criteria to determine potential airfield locations in one region of the country. Automate your site selection process using ModelBuilder and test your model in another part of the same country. Prepare a report with maps showing your recommendations for the new runway.

What to turn in

If you are working in a classroom setting with an instructor, submit an electronic copy or printout of your recommendations for the new runway.

Web resources

Feinstein International Center; Tufts University, Medford, Mass.: https://wikis.uit.tufts.edu/confluence/display/FIC/Movement+on+the+Margins.

Forced Migration Review: Refugee Studies Center, University of Oxford, England: http://www.fmreview.org.

IRIN Humanitarian News and Analysis. United Nations Office for the Coordination of Humanitarian Affairs: http://www.irinnews.org/Africa-Country.aspx?Country=UG.

Shuttle Radar Topography Mission (SRTM) Web site: http://www2.jpl.nasa.gov/srtm.

Uganda Humanitarian Clusters: http://www.ugandaclusters.ug/.

Uganda: Uncertain future for IDPs while peace remains elusive (a profile of the internal displacement situation) April 24, 2008: http://www.internal-displacement.org/.

Chapter 9

Improving the design and operation of refugee camps

The humanitarian imperative is to prevent and alleviate human suffering, protect life and health, and ensure respect for the human being. In committing to it, the emergency management community's intention is to support international humanitarian law, human rights, and the dignity of all people.

In this chapter, you will continue your study in northern Uganda by applying spatial analysis to improve the quality of humanitarian services at an actual IDP camp called Padibe. First, you will use satellite imagery to map camp infrastructure and assess population density. Then you will apply geostatistical analysis to improve water accessibility and assess Padibe's compliance with good practices in camp design.

Humanitarian guidelines to improve quality of service

In the previous chapter, you identified potential sites for a new IDP camp in northern Uganda. In this chapter, your goal is to optimize services and distribution of infrastructure within an actual camp near the Sudanese border, called Padibe. You will assess its current layout and then propose some design changes to improve the overall quality of life for camp inhabitants.

The analytical methods used in this lesson demonstrate the utility of spatial analysis in promoting good practices during humanitarian operations. Popular sources of information about these good practices include the following:

- Sphere Humanitarian Charter and Minimum Standards in Disaster Response (http://www.sphereproject.org)
- UNHCR *Handbook for Emergencies* (http://www.unhcr.org; go to Resources > Publications)
- USAID *Field Operations Guide for Disaster Assessment & Response* (http://www.usaid.gov)

All of these documents are available for free download from the Web sites given. Before starting this chapter, familiarize yourself with these quality-of-service guidelines and pay particular attention to the quantitative indicators that relate to camp management. You may wish to concentrate on the sections entitled "Shelter, Settlement & Non-Food Items" (chapter 4 of the Sphere Handbook, 2004); "Site Selection, Planning & Shelter" (chapter 12 of the UNHCR Handbook, 2007); and "Shelter & Settlements" (section F of the USAID Field Operations Guide, version 4.0).

Geostatistics

Although statistics can be difficult to understand and unattractive to decision makers, they are a powerful means to simplify large datasets that would otherwise remain unintelligible. Geostatistics measure geographic distributions, identify patterns and clusters, and analyze geographic relationships. Measuring geographic distribution can reveal the center, compactness, and orientation of spatial features. For example, determining the spatial mean, median, and mode for a set of points produces centers, which are much more comprehensible points.

Toward the end of this chapter you will perform two types of statistical analysis: mean center analysis and Voronoi analysis. A **mean center**, or centroid, is the weighted or unweighted balance-point for a set of points. Centroid analysis is particularly helpful in identifying trends of various phenomena over space and time.

Voronoi analysis calculates the area surrounding any point that is closer to that point than to any other point, called **Voronoi polygons**. This form of analysis can be a very powerful way to define service areas within a camp or settlement, and to optimize the relief distribution.

While it is always wise to be careful when generating and interpreting statistics, there is no need to be intimidated by them. You will see how they can be an important dimension of humanitarian GIS.

Scenario: Optimizing services at Padibe camp

As you learned in chapter 8, civil war in northern Uganda has displaced as many as two million people—about 80 percent of them women and children—over the past 20 years. Many live in camps that were established so long ago that they are no longer able to effectively service their inhabitants.

Even once camps have been established, however, it is not too late to enhance their design and operations in accordance with guidelines provided by the Sphere Charter, UNHCR Handbook, and others. Using the case study of Padibe camp, you will analyze and then optimize humanitarian infrastructure within this actual IDP camp in northern Uganda.

In exercise 9.1, you will map the camp in its present configuration using a high-resolution satellite image and tabular data. Then you will assess the camp's population density in exercise 9.2 to determine its compliance with the Sphere and UNHCR guidelines. Finally, in exercise 9.3, you will apply geostatistics to improve water services throughout the camp, in accordance with UNHCR guidelines.

Padibe camp is located about 400 kilometers north of Uganda's capitol, Kampala, about 30 kilometers from the Sudanese border. The camp was built in February 1997 after the first wave of IDPs arrived in flight from the Lord's Resistance Army attack on the Acholi people of northern Uganda.

Padibe camp is actually composed of two separate camps, Padibe East and Padibe West. Their combined population is about 30,000 IDPs, fairly typical of camps in the region. Like other camps, the operation of Padibe is segmented by humanitarian cluster. A humanitarian cluster is a group of agencies, interconnected by their respective mandates (such as health, education, or sanitation), that try to coordinate their work to improve the overall effectiveness, efficiency, and accountability of a humanitarian intervention.

The following table lists organizations involved in camp operations as of November 2009. (More details about these organizations and the clusters approach to humanitarian assistance, as well as maps and geodata, are available from the UN's regional Web site http://www.ugandaclusters.ug.)

Education	AVSI, FHU, CKS, WCH, CARITAS, STF/TT, KICWA, IRC, UNICEF
Food Security and Agricultural Livelihoods	FAO, KIDFA, STF/TT, FHU, WFP, NACCRI,CKS, KICWA
Governance, Infrastructure, and Livelihoods	FHU, UNDP, ACET, FAO, WCH, IRC, STF/TT
Health, Nutrition, and HIV/AIDS	AVSI, MP, FHU, WHO, IRC, AMREF, URCS, CKS
Protection (Human Rights, Rule of Law, Child, GBV, Conflict Mitigation, and Management)	OHCHR, UNICEF, WCH, ACDA, USAID/NUTI, IRC, FHU, UNFPA, IMC
Water, Sanitation, and Hygiene	FHU, OXFAM-GB, UNICEF

Table 9.1   Camp management agencies by cluster
(Source: UN OCHA November 2009)

Layer or attribute	Description
All_Huts	**Padibe Camp hut points (feature class)**
PPCODE	Padibe Camp place code (east and west)
BLOCK_NAME	Padibe Camp block name
HUT_ID	Hut identification number
All_Boreholes	**Padibe Camp borehole points (feature class)**
STATUS	Borehole service status (existing or potential)
Latrines	**Padibe Camp latrine points (feature class)**
Water_Point.dbf	**Padibe Camp water point locations (table)**
LATITUDE	Latitude coordinate in decimal degrees
LONGITUDE	Longitude coordinate in decimal degrees
Padibe_Image.tiff	**Padibe Camp IKONOS-2 1m Pan/4m MS satellite image**

Table 9.2   Data dictionary

Exercise 9.1

Mapping camp infrastructure

One of the first steps in mapping a camp is to locate its perimeter and subdivisions. Camps that are purpose-designed are usually organized with three administrative levels—zone, block, and plot. In the case of northern Uganda, however, most camps were not planned and exhibit a more random layout.

Create vector data from satellite (raster) imagery

The internal boundaries and perimeter of a camp can be collected several different ways: they can be recorded by walking their length with the tracking function of a GPS; they can be traced from a georeferenced sketch drawn by an agency active in the camp; and they can be extracted from remotely sensed imagery captured by aircraft or earth observation satellite.

When using remotely sensed data, it is important to "truth" your extracted features by making site visits or seeking input from colleagues familiar with the location. In this exercise, you will use data collected by field staff who know the camp well to construct your camp's spatial data infrastructure.

To display a camp border layer, first you need to create a blank polygon shapefile. This can be done in ArcCatalog.

1. **Start ArcCatalog by double-clicking the ArcCatalog icon. Make a connection to the Chapter9 folder within the GISHUM folder.**

2. **Right-click the Chapter9 folder, and then navigate to the Padibe camp geodatabase's Administrative features dataset.**

3. **Create a new feature class.**

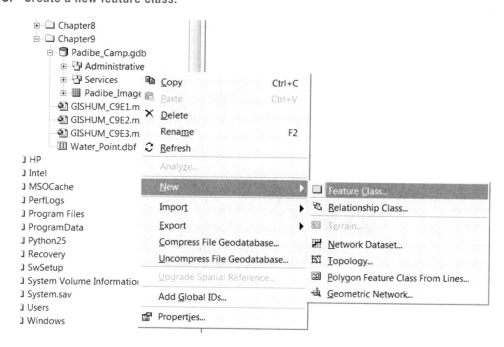

4. Name the new feature class **Padibe_Blocks** and assign polygon features as its type. Accept all other defaults by clicking Next, Next, and then Finish.

You have now generated a new (empty) layer with which to digitize the Padibe's seven camp blocks.

5. Minimize the ArcCatalog window, and then launch ArcMap. From the Chapter9 folder, open GISHUM_C9E1.mxd.

6. You will see the following satellite image: Padibe_Image. Use the Add Data button to add the new Padibe_Blocks layer from your camp geodatabase.

To add data to the Padibe_Blocks layer, you must first start an editing session.

7. Open the Editor Toolbar by going to View > Toolbars > Editor.

8. From the Editor drop-down list, select Start Editing. The Editor toolbar indicates that you will be creating a new feature in the Padibe_Blocks shapefile.

You are now ready to digitize the layout of the camp blocks within Padibe camp. Use the following graphic as a rough guide for creating seven blocks within the camp. In most cases, the blocks are separated by roads, so as you digitize you may wish to zoom in just enough to distinguish primary and secondary roads within the camp.

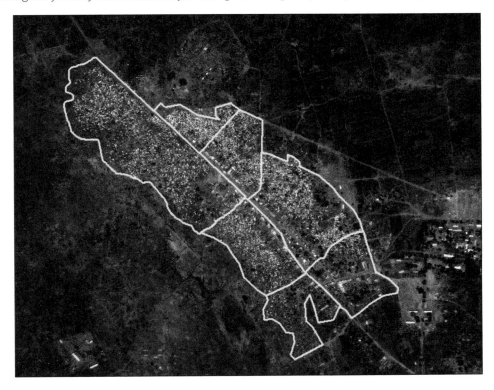

9. On the Editor toolbar, click the Pencil tool to begin the sketching process.

10. Begin tracing around the border of your first block. When you have traced the first block completely, right-click it, and then choose Finish-Sketch.

11. Before tracing the polygon for the next block, choose the Auto-Complete Polygon option from the Task drop-down menu on the Editor toolbar.

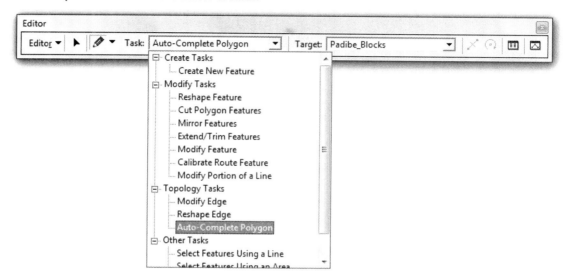

The Auto-Complete Polygon function allows you to create adjoining polygons without having to retrace the boundary between them. The polygons should share a border, but you want to avoid digitizing the border twice or having overlaps or spaces between polygons.

12. Once again, click the Pencil tool. Begin sketching the second block from inside the boundary of the first block. (For the Auto-Complete Polygon tool to be effective, the new sketch must cross or intersect the existing polygon edge at least two times for the new polygon to be created.)

13. Continue to trace the border of the second block. To finish, click once inside the boundary of the first block.

The following illustrates the work of the Auto-Complete Polygon tool (the satellite image has been turned off for ease of viewing). The dangling vertices will be automatically deleted once you click to finish the sketch for the second block.

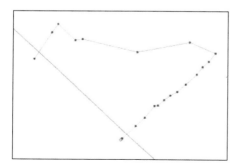
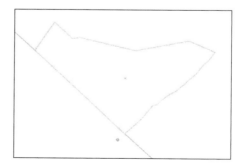

14. After the remaining blocks have been digitized, click Stop Editing from the Editor drop-down menu.

15. You are now ready to name each feature in the Padibe_Blocks layer. Open its attribute table, go to Options > Add Field, and create a new text field called **Name**.

16. Using the names of each block provided by the following graphic, populate the Name attribute of each block by starting a new editing session and directly naming each feature.

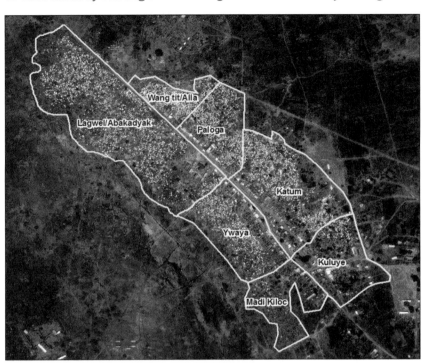

17. Right-click each block, and then select Attributes to open that features attribute information.

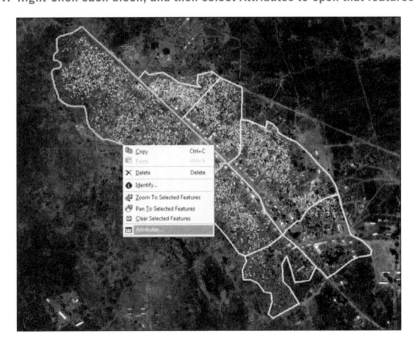

18. Type in the name of the feature using the reference map shown in step 16.

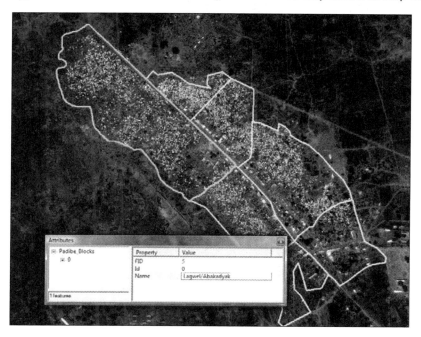

19. Once the remaining blocks have been named, click Stop Editing from the Editor drop-down menu.

20. Return to ArcCatalog, highlight the Padibe_Blocks layer, and click the Metadata tab. From the Metadata toolbar, choose FDGC ESRI as the Stylesheet option, and then click the Edit Metadata button.

21. Beginning with the General tab, complete all fields marked "REQUIRED" using the information you have (and do not have) about the data.

Your turn

Create a polyline feature class for Padibe's roads. Repeat steps 2 through 4 to create a new feature class within the Services feature dataset named **Padibe_Roads**. Then return to ArcMap and add the roads layer to your display. Start a new editing session, and then trace each of the roads and pathways that you can see on the image. You may need to zoom and pan across the image to accurately digitize the primary and secondary roads within the camp.

Once you have finished digitizing, create a new field in the Padibe_Roads attribute table called **Type**, and classify each road segment as either Primary, Secondary, or Tertiary. When you are finished creating the roads layer, stop the editing session and return to ArcCatalog to complete the essential metadata for your new data. (Revisit exercise 3.4 if it has been a while since you last created metadata in ArcCatalog.)

Your digitized roads layer will be similar to the following example, depending upon the precision and detail you employ. Use this exercise to test your skills in digitizing satellite imagery—very often the best source of geodata during humanitarian emergencies.

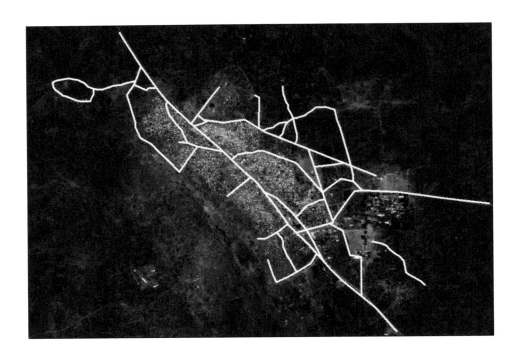

Convert tabular x,y data into vector data

Imagine that you have just received a table containing the GPS coordinates for water wells that are located throughout Padibe camp. In this part of the exercise, you will convert those locations from database file (*.dbf) format to shapefile (*.shp) format.

1. **Click Add Data.**

2. **From the Padibe camp geodatabase, add the Water Points table.**

3. **Open the table to see that each water point record contains fields displaying its latitude and longitude coordinates. You will use these two fields to generate a feature layer from the table.**

4. **Close the table.**

5. **Go to Tools on the main menu, and then click Add XY Data.**

6. **Water_Point.dbf is the only table in this ArcMap session, so it will appear by default. Ensure that the X and Y fields are correctly set so that X = Longitude and Y = Latitude. (This must be correctly assigned before proceeding further.)**

7. **The coordinate system is unknown (as yet). Click the Edit button. In the next window, click Select, and then choose the following path:**
       Geographic Coordinate Systems
           World
               Select the WGS 1984 file

8. Click Add. The Add XY Data wizard should resemble the following graphic.

9. Click OK. The water points have been added to the map window.

Notice that the name beside the symbol says Water_Point Events. This denotes that this layer is temporarily created for the current session only and not a permanent shapefile or feature class.

10. To create a permanent layer from the events theme, right-click Water_Points Events, scroll down to Data, and select the Export Data option.

11. Select to export the events layer into the Services feature dataset of the camp geodatabase as a new feature class called **WaterPoints**.

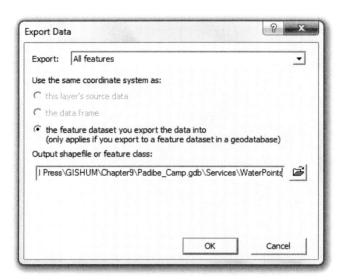

12. Click OK, and then replace the temporary events layer in your map with the geodatabase's WaterPoints layer.

13. Minimize the ArcMap window, and then return to ArcCatalog. Do your best to complete the required fields of metadata.

Obviously, you have very little knowledge about the source and method that was used to create the Water_Points database file in the Chapter9 dataset. That is a typical problem during humanitarian operations; your goal, therefore, is to disclose whatever you know—and what you don't know—when preparing metadata.

Exercise 9.2

Analyzing population density

Use count points per polygon to determine hut count by administrative zone

You have already observed that it is possible to distinguish individual huts in the satellite image of the camp. Given an average household size (persons per hut), the image allows the skilled GIS professional to analyze population density in different parts of the camp in order to offer camp managers guidance in how to improve its layout and infrastructure.

In this exercise, you will measure population density by calculating the number of huts in each subdivision of Padibe camp. You will see that significant differences in population density exist within the camp, a disparity that is bound to affect the quality of life and patterns of movement of its inhabitants.

1. Open a new ArcMap session. From the Chapter9 folder, open the project file GISHUM_C9E2.mxd.

2. Create a field called **Count** in the attribute table of the All_Huts layer, and then set its type to Short Integer.

3. Calculate the Count field equal to 1 by right-clicking the field name and then selecting Field Calculator. Type **1** in the white dialog box, and then click OK.

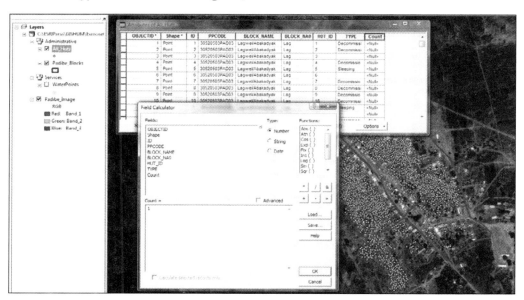

You have now assigned each hut record a count value of 1. Next, you will create a spatial join between your points layer and your polygon layer.

4. Right-click the Padibe_Blocks layer, and then select Joins and Relates > Join. From the drop-down list, choose "Join data from another layer based on spatial location."

5. **Populate the Join Data wizard as follows:**

   1. Chose the layer to join: All_Huts.
   2. How do you want the attributes to be summarized?: Sum.
   3. Save your output layer as a feature class, under the Administrative feature dataset of your geodatabase, and name the output layer **HutsPerBlock**.

6. **When your settings appear as follows, click OK.**

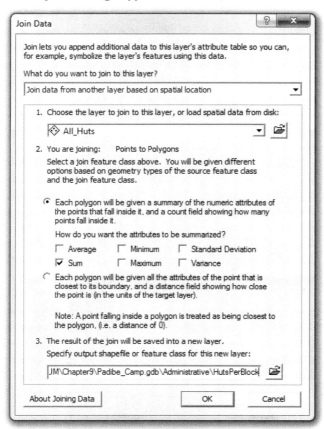

Once the join has been calculated, the new HutsPerBlock layer will be added to your map display, and will include a field with a total count of the huts contained within each block in Padibe camp.

7. **To calculate the hut density, add a new field called Area, and then set its type to Float.**

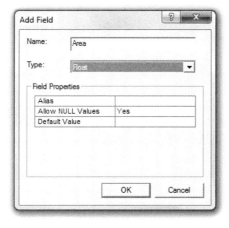

8. Right-click the new field to calculate each block's area in hectares.

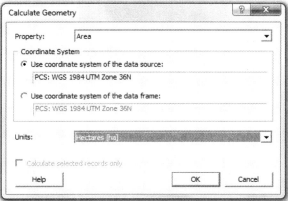

9. Now add another new field called **Density**, and use the Field Calculator to compute the number of huts per hectare for each block.

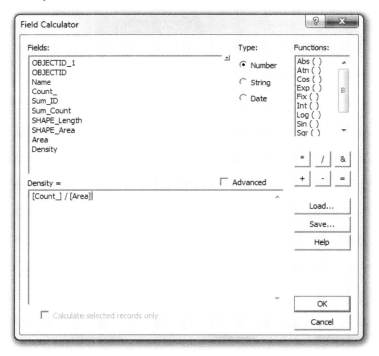

10. Your HutsPerBlock attribute table now provides you with a crude measure of population density (average huts per hectare for each block). Once you have turned off unnecessary fields in your table, your analysis can be sorted according to the density.

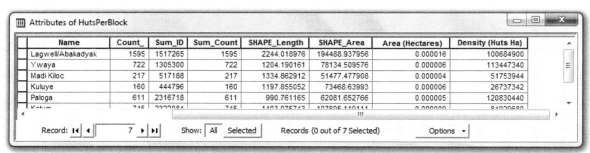

Name	Count_	Sum_ID	Sum_Count	SHAPE_Length	SHAPE_Area	Area (Hectares)	Density (Huts Ha)
Lagwell/Abakadyak	1595	1517265	1595	2244.018976	194488.937956	0.000016	100684900
Ywaya	722	1305300	722	1204.190161	78134.509576	0.000006	113447340
Madi Kiloc	217	517188	217	1334.862912	51477.477908	0.000004	51753944
Kuluye	160	444796	160	1197.855052	73468.63993	0.000006	26737342
Paloga	611	2316718	611	990.761165	62081.652766	0.000005	120830440
Katum	745	2322084	745	1403.075743	107605.110111	0.000000	84020680

Record: 14 ◄ 7 ► ►I   Show: All | Selected   Records (0 out of 7 Selected)   Options ▾

(Because your digitized block polygons undoubtedly differ from the author's, your values for Sum_Count, Area, and Density will also differ. Take this as a lesson in the consequences of digitization—and remember that the quality of your feature digitization will significantly affect your ultimate analysis.)

11. **Use this information to create a thematic map of hut density. Customize and then save your map as a PDF file in your Chapter9 data folder.**

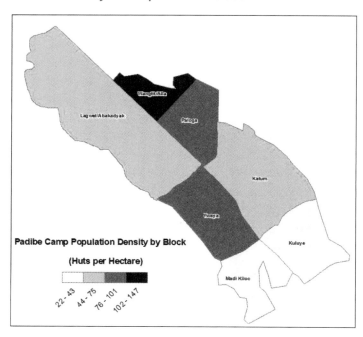

Create a raster layer to map hut density

ArcGIS Spatial Analyst extension provides an alternative method of mapping population within the camp, called density analysis. The density function distributes a measured quantity of an input point layer throughout a landscape to produce a continuous surface.

Available density mapping tools include the following:

- Kernel Density
- Line Density
- Point Density

In the case of Padibe camp, the density analysis raster output can be used to communicate the congestion associated with the actual location of huts, and the trends in hut locations over time (understanding the "spread" of features). Note that when the values of all the cells in a density layer are added up, they equal the sum of the population of the original point layer.

1. **In ArcToolbox, navigate to Spatial Analyst Tools > Density > Point Density.**

2. **Populate the Point Density Wizard as follows:**

   - Input point features: All_Huts
   - Population field: NONE
   - Output raster: Chapter9\Padibe_Camp.gdb\HutDensity
   - Output cell size: 20
   - Neighborhood: Circle
   - Area units: HECTARES

3. **Click OK.**

4. **Resymbolize your raster layer using a defined interval of 25 huts per hectare.**

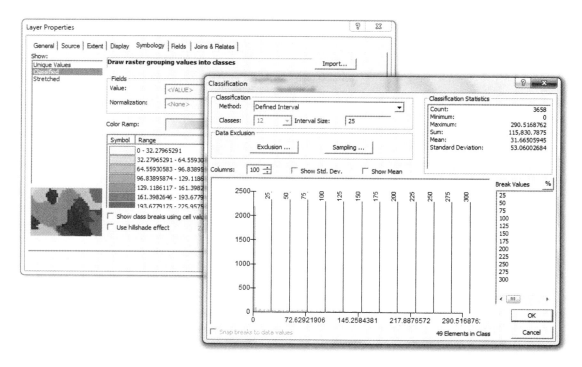

The block boundaries that you digitized using the satellite image were too crude to accurately reflect the camp's population density. In comparison, the HutDensity layer measures population density at the subblock level, producing a much more reliable assessment of population density throughout Padibe camp.

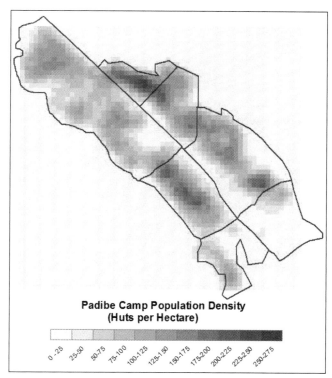

**Padibe Camp Population Density
(Huts per Hectare)**

0 - 25 25-50 50-75 75-100 100-125 125-150 150-175 175-200 200-225 225-250 250-275

Measure adequacy of camp area using Sphere and UNHCR guidelines

In your reading assignment for this chapter, you learned that the humanitarian community has attempted to set quality-of-service guidelines for shelter planning related to space requirements. The following table summarizes the spatially relevant standards provided by the Sphere Project and the UNHCR *Handbook for Emergencies*:

Space required	The Sphere Project (2004)	UNHCR (2007)
Minimum surface area of camp per person	45 m^2, including infrastructure (pp.216–217).	45 m^2 per person recommended (including garden). Should not be less than 30 m^2 per person (p.210). 30–45 m^2 per person (p.549).
Minimum covered floor area per person	At least 3.5 m^2, except in extreme circumstances (pp. 219–220). 3.5 m^2 in warm climate.	4.5–5.5m^2 in cold climate or urban situations, including kitchen and bathing facilities (p.221).

Table 9.3

The guidelines given in the table intend to afford displaced populations with dignified levels of personal privacy and opportunity to improve their livelihoods through microscale farming and industry. Their prospects of eventually returning to normal life depend largely on whether displaced people have the chance to pursue vocational trades such as metalwork, arts and crafts, and so on, while still in the camp.

You will now resymbolize your population density map for Padibe camp in order to assess how it compares with the minimum surface area per person offered by the above standards. For simplicity, assume that every hut in the camp represents four people. (Just keep in mind, though, that the Count field in the previous section would enable you to perform your analysis if you had actual household statistics for each hut in the camp.)

1.   **Right-click your HutsPerHectare layer, and then select Properties.**

You will now manually classify the range break values of the layer to reflect the following conversion formula: Square Meters/Person = 1/ (Huts/Hectare * 4 Persons/Hut * 1 Hectare/10,000 Square Meters)

Using this formula, it is easy to calculate the range breaks that better reflect the Sphere Project's and UNHCR's guidelines:

Square Meters/Person	Huts/Hectare
7.5	333.3
10	250.0
15	167.6
22.5	111.1
30	83.3
45	55.6
60	41.2

Table 9.4

2. **Manually reclassify the symbology break values that reflect the ranges shown in the left column.**

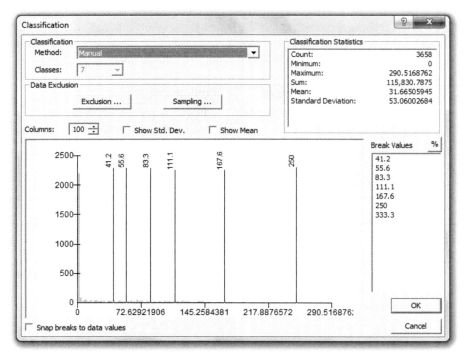

3. **Click OK.**

4. **Select the green-to-red color ramp, and then click the Range column title to reverse sorting (since you have inverted the values by converting from Huts/Hectare to Square Meters/Person).**

5. **Now type the given conversion values directly into the label column so that they reflect the actual range values in Square Meters/Person (as opposed to Huts/Hectare, which is actually what they are).**

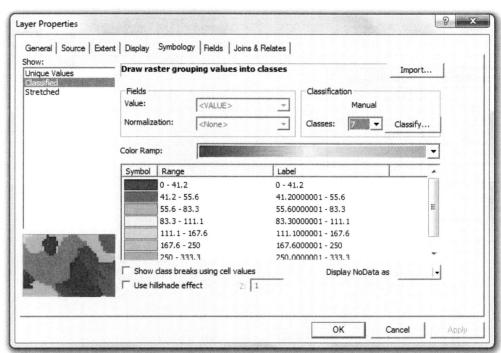

6. Regenerate your Padibe camp Population Density map in m²/person; label the camp blocks; and in layout view, add a legend and scale bar. You may also wish to add additional information or apply a map template that reflects your organization's cartographic standards.

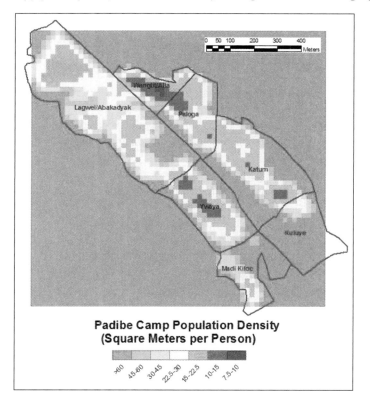

**Padibe Camp Population Density
(Square Meters per Person)**

7. Right-click the HutsPerHectare layer, and then select Save as Layer File. Name the layer **IDP_Space_Analysis.lyr**, and then save it to your Chapter9 data folder.

8. Add the new IDP_Space_Analysis.lyr to your map display, and then open its layer properties. Under the General tab, rename the layer **IDP Space Analysis (sqm/person)**.

What to turn in

If you are working in a classroom setting with an instructor, turn in a printed map, screen capture, or PDF file of your camp population density map in square meters per person.

Produce a unique address for each hut using Field Calculator

Before completing our study of huts in Padibe camp, you will assign each hut a unique address. An address system in a camp is very useful for communicating with its inhabitants, delivering mail and supplies, and providing special assistance to certain households. An address system also enables inhabitants to be associated with camp infrastructure, a very powerful means to assess the quality of humanitarian services as camp size and complexity increases.

After consultation with stakeholders, you have decided that the address shall be composed of three fields:

- Place code: Place codes (P-codes) are unique alphanumeric codes that identify village or location. P-codes have been used to reduce the confusion arising from the multilingual, multicultural, multiagency environments that typically exist during humanitarian emergencies.
- Block code: For Padibe camp, you have decided to use the first three letters of the actual name of each block.
- Household number: For Padibe camp, you are going to use the Hut ID numbers to represent each household.

In the last part of this exercise, you will merge the three fields into a single field that geocodes every hut with a unique address. The United Nations Regional Office for Central and East Africa has given Padibe camp the P-codes 30520503PAD02 and 30520503PAD03. You will combine those P-codes with an abbreviation for each block and a simple number for each hut to create an alphanumeric string, called Address.

1. **Open the attribute table for the All_Huts layer.**

2. **Click Options > Add Field.**

3. **Name the new field Address, and set its type as Text and its length as 50.**

4. **Right-click the new Address field, and then navigate to Field Calculator.**

5. **Generate the new Address value using the expression in the following image.**

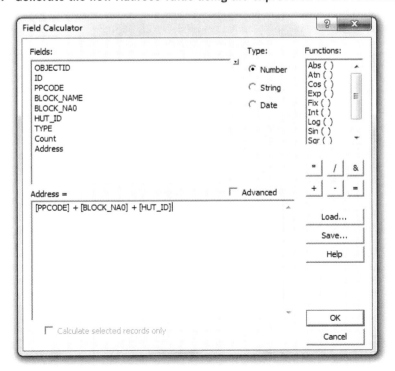

6. Click OK. Your All_Huts attribute table now includes a unique identifier for each hut record, which can serve as a simple, georeferenced address for every household in Padibe camp.

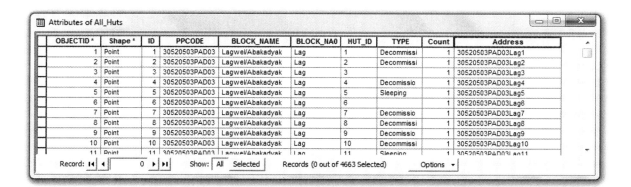

OBJECTID*	Shape*	ID	PPCODE	BLOCK_NAME	BLOCK_NA0	HUT_ID	TYPE	Count	Address
1	Point	1	30520503PAD03	Lagwel/Abakadyak	Lag	1	Decommissi	1	30520503PAD03Lag1
2	Point	2	30520503PAD03	Lagwel/Abakadyak	Lag	2	Decommissi	1	30520503PAD03Lag2
3	Point	3	30520503PAD03	Lagwel/Abakadyak	Lag	3		1	30520503PAD03Lag3
4	Point	4	30520503PAD03	Lagwel/Abakadyak	Lag	4	Decomissio	1	30520503PAD03Lag4
5	Point	5	30520503PAD03	Lagwel/Abakadyak	Lag	5	Sleeping	1	30520503PAD03Lag5
6	Point	6	30520503PAD03	Lagwel/Abakadyak	Lag	6		1	30520503PAD03Lag6
7	Point	7	30520503PAD03	Lagwel/Abakadyak	Lag	7	Decommissi	1	30520503PAD03Lag7
8	Point	8	30520503PAD03	Lagwel/Abakadyak	Lag	8	Decommissi	1	30520503PAD03Lag8
9	Point	9	30520503PAD03	Lagwel/Abakadyak	Lag	9	Decomissio	1	30520503PAD03Lag9
10	Point	10	30520503PAD03	Lagwel/Abakadyak	Lag	10	Decommissi	1	30520503PAD03Lag10
11	Point	11	30520503PAD03	Lagwel/Abakadyak	Lag	11	Sleeping	1	30520503PAD03Lag11

Record: 14 ◄ [0] ► ►I   Show: All Selected   Records (0 out of 4663 Selected)   Options ▼

Exercise 9.3

Applying humanitarian service guidelines

Now that you have mapped and studied the general character of the camp, you are ready to assess the quality of its services relative to humanitarian best practices.

In this exercise you will do the following:

- Assess compliance with quality-of-service standards for water distribution.
- Optimize the location of an additional water point within the camp.
- Determine if a disease outbreak in the camp might be associated with the layout of water and sanitation facilities.
- Create firebreaks (safety corridors) in the camp with a minimum amount of disruption to existing inhabitants.

Review the table you created at the beginning of this lesson; it summarizes the quantitative guidelines associated with the provision of shelter services.

Improve water accessibility using UNHCR Handbook for Emergencies

Access to sufficient, clean water is often a challenge for IDPs and refugees, who might spend many hours each day fetching water from distant and unreliable sources. Not only does this expose women and children to risk, it is also a less productive use of time than other activities such as cooking, child care, basic education, and vocational training.

The humanitarian community has attempted to set quality-of-service guidelines for shelter planning related to water supply. The following table summarizes the spatially relevant standards provided by the Sphere Project and the UNHCR *Handbook for Emergencies*.

Water supply	The Sphere Project (2004)	UNHCR Handbook (2007)
Minimum quantity of water available (liters per person per day)	15 (p.63)	15–20 (p.549)
People per tap-stand	Maximum 250 (p.65)	1 tap per 200 people not farther than 100 meters from user accommodations (p.549)
Distance from dwellings to taps	Maximum 500 meters (p.63)	Maximum 100 meters or a few minutes' walk (p.219)

Table 9.5

Your organization aspires to reach the UNHCR's standard for water accessibility at Padibe camp and would like to ensure that huts are no farther than 100 meters from water taps (hereon referred to as "water points," each of which includes one or more individual tap stands).

1. **Open GISHUM_C9E3.mxd. You should see the location of existing water points in relation to the All_Huts and Padibe_Block layers.**

2. Start ArcToolbox, and then choose Analysis > Proximity > Buffer.

3. Populate the buffer wizard as follows in order to create a 100-meter buffer around the water points:

   - Input Features: WaterPoints
   - Output Features: Chapter9\Padibe_Camp.gdb\Services\WaterPoints_Buffer
   - Linear unit: 100 Meters
   - Dissolve Type: ALL

4. Click OK. The buffer reveals that many huts in Padibe camp are farther than 100 meters from a water point.

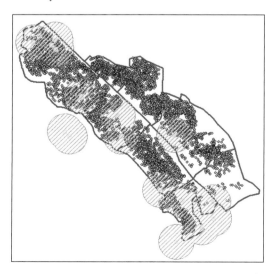

You will now calculate the proportion of huts that meet the UNHCR's guidelines for water accessibility.

5.   **Go to Selection > Select By Location.**

6.   **Populate the wizard to select features from All_Huts that are completely within the WaterPoints_ Buffer layer.**

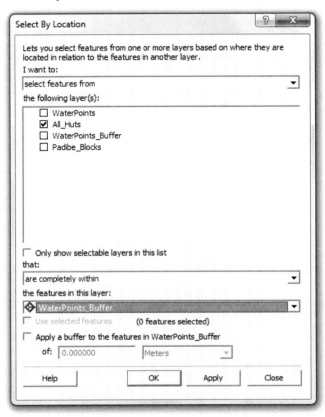

7.   **Click OK. The Huts that fall within the 100-meter buffer zone will be highlighted in the map.**

8.   **To save these huts as a separate shapefile, right-click the All_Huts layer, navigate to Data > Export data, and add the new layer to the Services feature dataset in your camp geodatabase. Name it HutsNearWater.**

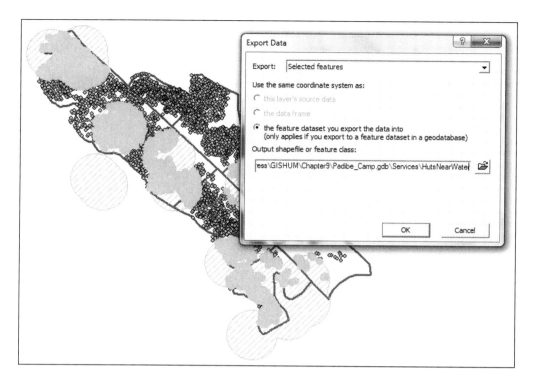

9. Click OK to export the selected hut data, and select Yes if asked if you want to add the new layer to the map.

10. To create a separate shapefile of Huts that are outside the 100 buffer area, right-click the All_Huts layer, and click Selection > Switch Selection. (If you have cleared the selection of huts within the 100-meter buffer, reselect them using the Select By Location tool).

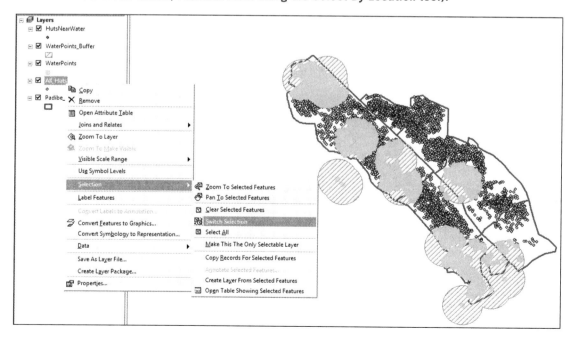

11. **Click Yes to continue when warned that the Switch Selection operation could take a long period of time if there are many records in the table.**

12. **Now all huts outside the 100-meter buffer are selected. Export this selection to the Services dataset as HutsFarWater and add the new layer to the map.**

13. **Open the attribute tables of the HutsNearWater and HutsFarWater layers to determine the actual number of huts (total number of records) within and beyond the 100-meter water accessibility guidelines. Note the following:**

    1. Number and percentage of huts that have adequate water access:_____
    2. Number and percentage of huts that do not:_____

Because more than half of the camp population does not have satisfactory water access, donors agree to reduce hardship in the camp by funding improved water accessibility. They will support the following:

- Drilling of two new water points anywhere in the camp
- Relocation of two existing water points (note that extension pipelines cannot cross any block boundaries; in other words, the water point can only be relocated within its existing block, not onto another block)

Your mission in the next part of the exercise is to recommend the locations for these optimized water points and to calculate the improvement in water access throughout the camp.

Optimize water services using mean center analysis

One of the most common applications of geostatistics is mean center analysis, which calculates the mean average of a set of points. That single point reflects the weighted or unweighted average location of the points. In the case of Padibe camp, calculating the mean center of the huts in each block will help determine the optimal location to position water points for each camp block.

1. **Open the attribute table of your Padibe_Blocks layer, and then click the gray bar to the left of the record for Madi Kiloc block.**

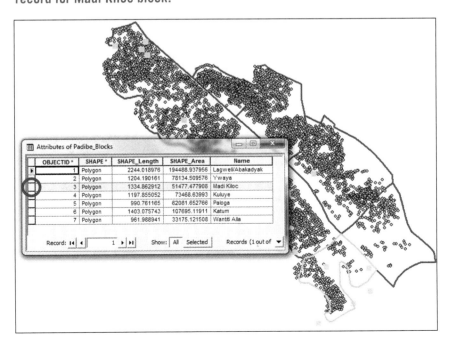

2. From the main menu, choose Selection > Select By Location. Populate the wizard to select features from All_Huts that are completely within the Padibe_Blocks layer. Check the "Use selected features" check box.

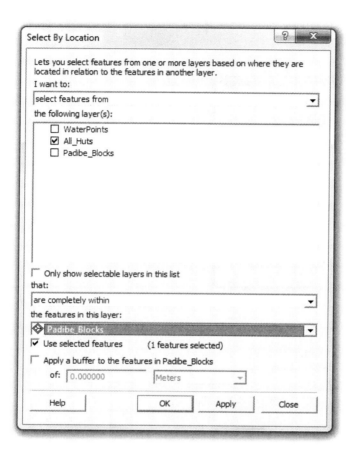

3. Click OK. You have now selected all of the huts that fall within Madi Kiloc block.

4. Start ArcToolbox.

5. Go to Spatial Statistics Tools > Measuring Geographic Distributions > Mean Center.

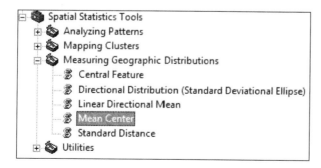

6. **Populate the wizard as follows to calculate the unweighted mean center of all huts in Madi Kiloc block:**

- Input Feature Class: All_Huts.
- Output Feature Class: Padibe_Camp.gdb\Administrative\Madi_MeanCenter.
- Leave all other fields blank.

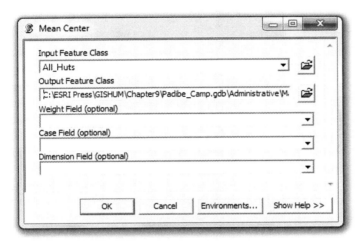

7. **Click OK.**

You should now see the mean center of all huts within Madi Kiloc block.

8. **Repeat steps 1 through 7 on pages 329 through 331 for the other camp blocks to create a total of seven mean center feature classes in your camp geodatabase.**

9. **In ArcToolbox, go to Data Management Tools > General > Merge.**

10. **Populate the Merge tool as follows to combine the seven separate mean center layers into a single layer:**

- Input Datasets: Madi_Huts_MeanCenter, plus the six other mean centers you created
- Output Feature Class: Padibe_Camp.gdb\Administrative\All_Huts_MeanCenter_Merge

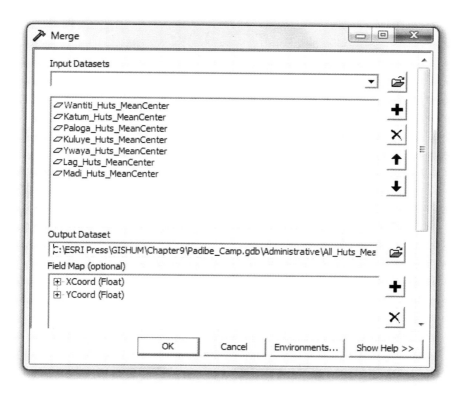

11. **Click OK.**

12. **Remove the separate mean center layers, and then resymbolize the merged layer to ensure it can be distinguished from the other layers displayed in your map.**

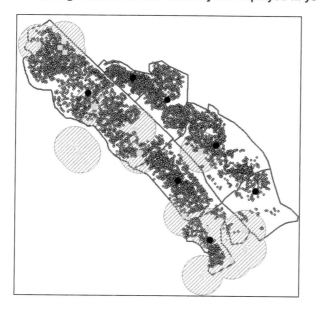

You are now in an informed position to assess what to recommend to optimize water points in Padibe camp in accordance with your donors' offer.

First, create a copy of your water points layer that can be used for analysis.

13. **Right-click the WaterPoints layer, and then select Data > Export Data. Save the layer to the Services feature dataset as OptimalWaterPoints.**

14. **Click Yes to add the new layer to your display, and then turn off the original WaterPoints layer.**

You are now ready to add two new water points. The most obvious gaps in water accessibility are in Wangtit/Alla and Pagola blocks, but relocation of an existing water point is not an option in either of those blocks. You will therefore assess the impact of drilling new boreholes at the mean center of the huts in each of those blocks.

15. **From the Editor drop-down menu, choose Start Editing. You may need to click View > Toolbars on the main menu if you cannot see the Editor toolbar.**

16. **Ensure that your task is set to Create New Feature and your target is OptimalWaterPoints.**

17. **Select the Sketch** 🖉 **tool, and then click in the mean center of the Pagola huts.**

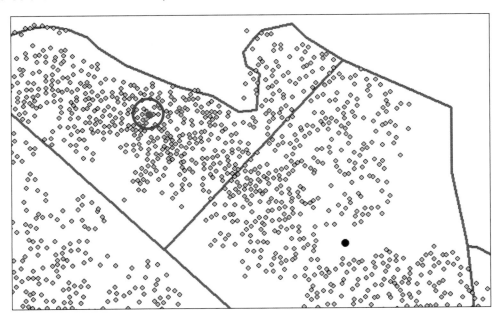

18. A new water point symbol will appear over the mean center symbol. Add another water point over the mean center of Wangtit/Alla's huts.

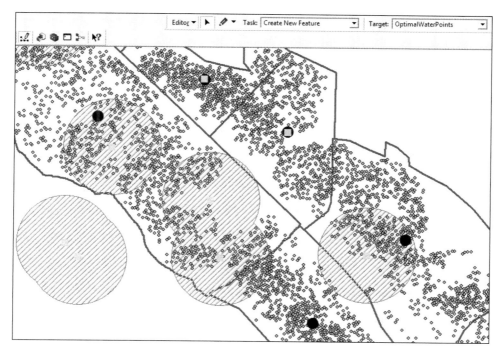

19. Now study the rest of the camp blocks and decide which two existing water points could be better situated within their current blocks.

20. Change your Editor task to Modify feature.

21. Click an existing water point symbol, and then drag the activated symbol to its new location. Click outside of the camp to show the new water point location.

22. Repeat for another existing water point, remembering that you cannot move more than two points and that they must remain within their current blocks.

23. Once you are happy with your repositioning of water points, open the attribute table of the OptimalWaterPoints layer, and then examine the Longitude (X) and Latitude (Y) values. You will now update the x,y coordinates of your optimized water points.

24. Right-click the Longitude field header, and then select Calculate Geometry. Populate the wizard as follows in order to calculate the longitude coordinates of each water point:

- Property: X Coordinate of Point
- Coordinate System: Use coordinate system of the data source
- Units: Decimal Degrees

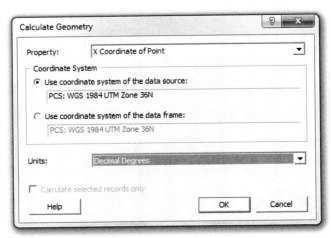

25. Click OK. Now repeat the process for the Latitude field using the property Y Coordinate of Point.

26. Assign the two new water points the following IDs: 2000 and 2010.

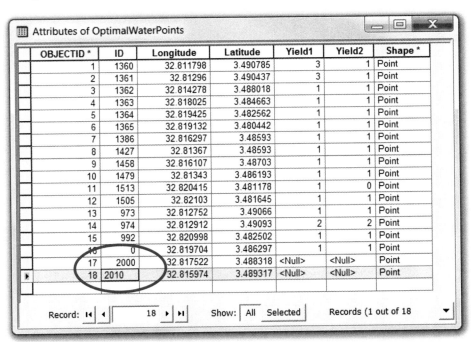

OBJECTID *	ID	Longitude	Latitude	Yield1	Yield2	Shape *
1	1360	32.811798	3.490785	3	1	Point
2	1361	32.81296	3.490437	3	1	Point
3	1362	32.814278	3.488018	1	1	Point
4	1363	32.818025	3.484663	1	1	Point
5	1364	32.819425	3.482562	1	1	Point
6	1365	32.819132	3.480442	1	1	Point
7	1386	32.816297	3.48593	1	1	Point
8	1427	32.81367	3.48593	1	1	Point
9	1458	32.816107	3.48703	1	1	Point
10	1479	32.81343	3.486193	1	1	Point
11	1513	32.820415	3.481178	1	0	Point
12	1505	32.82103	3.481645	1	1	Point
13	973	32.812752	3.49066	1	1	Point
14	974	32.812912	3.49093	2	2	Point
15	992	32.820998	3.482502	1	1	Point
16	0	32.819704	3.486297	1	1	Point
17	2000	32.817522	3.488318	<Null>	<Null>	Point
18	2010	32.815974	3.489317	<Null>	<Null>	Point

Record: |◄ ◄ | 18 | ► ►| Show: All Selected Records (1 out of 18)

27. From the Editor drop-down menu, choose Stop editing.

28. Click Yes to save your edits to the OptimalWaterPoints layer.

29. You are now ready to repeat steps 2 through 13 of this exercise to determine the impact of the new water point locations.

What to turn in

If you are working in a classroom setting with an instructor, prepare a short report illustrating the current and recommended water point locations that justifies your relocation proposal to the Water/Sanitation/Hygiene cluster leaders of the Padibe camp.

Calculate service populations using Voronoi analysis

The report you have just created provides guidance for determining where to drill new boreholes. While there is no guarantee that groundwater will be found at your two new water points, it is certainly logical to begin groundwater surveys at those locations.

The next steps in your water point optimization at the Padibe camp are to calculate the following:

- The number of taps required at each water point in the camp, based upon UNHCR Handbook's guideline of 1 tap per 200 people
- The total water production requirements for each water point, based upon the Handbook's minimum requirement of 15 liters per person per day.

Since these computations cannot be generalized at the block or camp level, you will apply a geostatistical technique called "Voronoi analysis" to divide the camp into service polygons that identify the closest huts around each water point.

1. From the Tools menu, navigate to Extensions, and then select Geostatistical Analyst.

2. From the View menu, navigate to Toolbars > Geostatistical Analyst.

3. From the Geostatistical Analyst drop-down menu (on the Geostatistical Analyst toolbar), choose Explore Data > Voronoi Map.

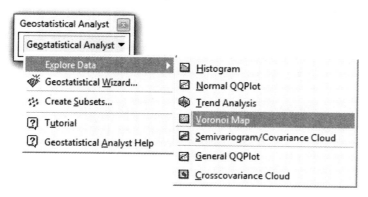

4. Populate the Voronoi Map wizard as follows in order to calculate the geostatistical service area for each water point:

   - Type: Simple
   - Clip Layer: None
   - Layer: OptimalWaterPoints
   - Attribute: ID

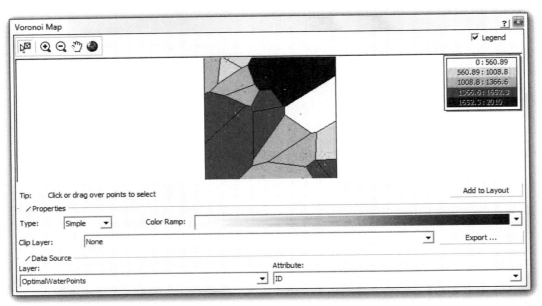

5. Click Export to save the Voronoi polygons as a feature class in your Services dataset of your camp geodatabase, and call it **OptimalWaterPointServiceAreas**.

6. Click Yes to add the exported data to your map, and then close or minimize the Voronoi Map wizard. Resymbolize the new layer with a blue outline color with a width of 2 (but no color fill). To truly appreciate the service area of each water point, you may want to limit your display to just the Padibe_Image, OptimalWaterPoints, and OptimalWaterPointServiceAreas layers.

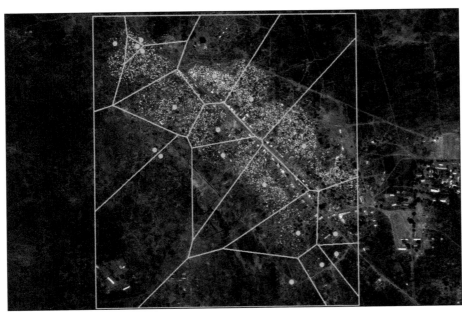

Your Voronoi polygons divide the camp by the service areas of your optimized water point locations. Realize that while you based your optimization of water accessibility using the mean centers for each camp block, these service area polygons disregard camp block boundaries. They simply define the closest water point for any hut in the Padibe camp—theoretically the most convenient water point for every household.

You are now ready to compute the number of taps and the water production for each water point.

7. **Right-click the OptimalWaterPointServiceAreas layer, and then select Joins and Relates > Join. Click the drop-down list, and then select "Join data from another layer based on spatial location."**

8. **Populate the Join Data wizard as follows:**

1. Chose the layer to join: All_Huts.
2. How do you want the attributes to be summarized?: Sum.
3. Save your output layer as a feature class under the Services feature dataset of your geodatabase.

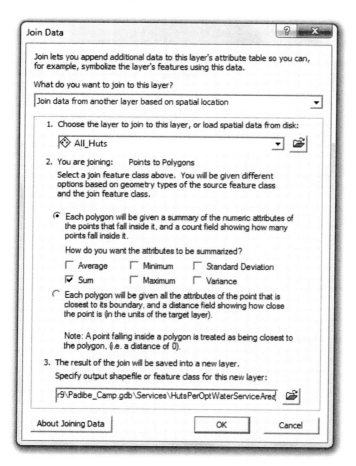

9. **Name the output layer HutsPerOptWaterServiceArea, and click OK.**

Once the join has been calculated, the new layer will be added to your map display and will include a field with a total count of the huts contained within each water point's service area.

10. **Open the attribute table of your HutsPerOptWaterServiceArea layer. Under the Sum_Count column you will see the number of huts closest to each water point than to any other.**

11. **Add three new long integer fields to the table, and then use the Field Calculator to compute their values as follows:**

PopnServed (alias: IDPs per ServiceArea)   since there are 4 persons per hut
Calculation: PopnServed = Sum_Count*4

MinTaps (alias: Minimum Taps Required)   since the UNHCR maximum is 200 persons per tap
Calculation: MinTaps = PopnServed/200

MinProdn (alias: Minimum Daily Water Supply)   since the UNHCR minimum is 15 liters per persons per day
Calculation: MinProdn = PopnServed*15

What to turn in

If you are working in a classroom setting with an instructor, prepare a map illustrating the service area polygons for your optimized water points with a table listing the population served, the minimum taps, and the minimum daily production for each water point to meet UNHCR guidelines.

The techniques taught in this chapter are powerful examples of how spatial statistics can be used to improve humanitarian operations at the camp level.

This could be the first time you've used GIS to perform statistical analysis and, depending upon your prior experience with statistics, you might find it helpful to study the theory and practice of the application of GIS to statistics. *The ESRI Guide to GIS Analysis, Volume 1: Geographic Patterns and Relationships* and *Volume 2: Spatial Measurements and Statistics*, by Andy Mitchell, are excellent references.

Assignment

Task A

Having conducted a groundwater survey of Padibe, hydrogeologists have provided you with a reliable map. It shows all existing and potential boreholes that can serve as water points throughout the camp. The Services feature dataset of our camp geodatabase includes those locations as a feature class called All_Boreholes.

Given available funding, you have two choices:

1. Add any two potential water points and relocate two existing water points, or
2. Add any three potential water points but make no relocations.

Using the All_Boreholes points layer (not your own OptimalWaterPoints layer), prepare recommendations. Include a map showing which existing and potential (new) water points you believe are optimal. Calculate the number of taps required at each water point using Voronoi analysis and the UNHCR's recommendation of 200 people per tap (assume 4 persons per hut throughout the camp).

Task B

Due to an outbreak of Hepatitis E in Kitgum District, you have been asked to prepare recommendations for improving the quality of sanitation offered throughout the camp. Produce a report detailing the least-cost method of enabling Padibe to achieve compliance with the sanitation guidelines of the Sphere Project. Use the Latrines feature class in the Services feature dataset of your camp geodatabase, and the following sanitation guidelines.

Sanitation	The Sphere Project (2004)	UNHCR (2007)
Maximum people per latrine	20 people (if sex-segregated public toilets) (pp.71–72)	In order of preference: (1) Family (5–10 people) (2) 20 people (p.549)
Distance from dwelling to toilet (sited to pose minimum threats to users, especially at night)	Maximum 50 meters (p.71)	6–50 meters (p.549)

Table 9.6

What to turn in

If you are working in a classroom setting with an instructor, turn in an electronic copy or printout of your recommendations.

A final word

By completing all nine chapters of this tutorial, you are now prepared for most routine and many advanced applications of humanitarian GIS. Take a moment to reflect upon how significantly your ArcGIS skills have improved since you began the first chapter—this book probably pushed you up a steeper learning curve than you originally envisioned, right?

While you should feel quite confident about your ability to perform as a humanitarian GIS specialist, don't forget to maintain and advance the skills that you've worked so hard to attain.

Appendix A

Coordinate systems

Maps are flat, but the surfaces they represent are curved. Transforming three-dimensional space into a two-dimensional map is called **projection.** Projection formulas are mathematical expressions that convert data from a geographical location (latitude and longitude) on a sphere or spheroid to a representative location on a flat surface. The process distorts at least one of the following properties—shape, area, distance, and direction—and often more. Because measurements of one or more of these distorted properties are often used to make decisions, anyone who uses maps as analytical tools should know which projections distort which properties, and to what extent.

The Projections and Transformations toolset in ArcMap contains tools to set a projection, to reproject datasets, or transform datasets. Properly projected data ensures accurate, reproducible GIS representations and measurements, and choosing the most appropriate projection is an essential data management skill for the humanitarian GIS professional.

All geographic datasets used in ArcGIS are assumed to have a well-defined coordinate system, which enables them to be located in relation to the earth's surface. If your datasets have a well-defined coordinate system, then ArcGIS can automatically integrate them with others by projecting your data on the fly into the baseline coordinate system.

If your datasets do not have a spatial reference, they may still be drawn in your ArcMap display, but they cannot be projected. You should therefore try to assign a spatial reference to your data layers before performing any analytical work such as measuring distances or areas.

Geographic coordinates versus projected coordinates

Features on a map reference the actual geographical locations of the objects they represent in the real world. The positions of objects on the earth's spherical surface are measured in geographic coordinates.

Geographic coordinates express location on the earth's surface in degrees of latitude and longitude (figure A-1). While latitude and longitude can specify exact positions on the spherical surface of the earth, they are not uniform units of measure. Only along the equator does the distance represented by one degree of longitude approximate the distance represented by one degree of latitude. To overcome measurement difficulties, data is often transformed from a three-dimensional geographic coordinate system to a two-dimensional projected coordinate system (see ArcGIS Desktop Help for additional background and details on the concepts discussed in this appendix).

A projected coordinate system defines locations on the earth's surface using a two-dimensional system that locates features based on their distance from an origin (0,0) along two axes: a horizontal x-axis representing east–west and a vertical y-axis representing north–south. Projected coordinates are transformed from latitude and longitude to x,y coordinates using a map projection.

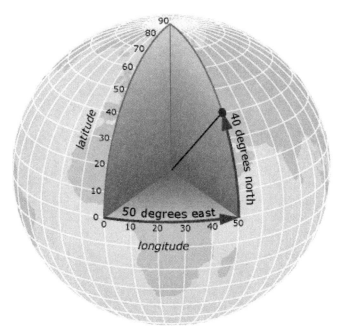

Figure A-1   Determining the geographic coordinates for a location 50° east longitude and 40° north latitude.

Spheroids and datums

While a spheroid approximates the shape of the earth, a datum defines the position of the spheroid relative to the center of the earth. A datum provides a frame of reference for measuring locations on the surface of the earth and defines the origin and orientation of latitude and longitude lines.

Whenever you change the datum, or geographic coordinate system, the coordinate values of your data will change. Since the 1990s, satellite data has provided scientists with new measurements to define the best earth-fitting spheroid, which relates coordinates to the earth's center of mass. An earth-centered, or geocentric, datum uses the earth's center of mass as the origin. The most widely used datum is WGS 1984. It serves as the framework for locational measurement worldwide, and for most of this book's datasets.

While WGS 1984 is popular worldwide, geodetic scientists and land surveyors often prefer to use local datums in order to more closely fit the earth's surface to a spheroid in a particular area. That level of precision is rarely needed for humanitarian GIS applications; however, you should be prepared to import data with very unfamiliar types of spatial reference into your geodatabase.

With local datums, a point on the surface of the spheroid is matched to a particular position on the surface of the earth. This point is known as the origin point of the datum. The coordinates of the origin point are fixed, and all other points are calculated from it.

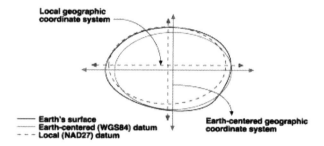

Figure A-2 Comparing local and earth-centered datums with the earth's actual shape.

The coordinate system origin of a local datum is not at the center of the earth (figure A-2). The center of the spheroid of a local datum is offset from the earth's center. NAD 1927 and the European Datum of 1950 (ED 1950) are local datums. NAD 1927 is designed to fit North America reasonably well, while ED 1950 was created for use in Europe. Because a local datum aligns its spheroid so closely to a particular area on the earth's surface, it's not suitable for use outside the area for which it was designed.

Map projections

One easy way to understand how map projections are created is to visualize shining a light through the earth onto a surface, called the projection surface (figure A-3). Imagine your model of the earth is clear glass with the graticule (latitudes and longitudes) drawn on it. Wrap a piece of paper around the earth. A light at the center of the earth will cast the shadows of the graticule onto the piece of paper. If you unwrap the paper and lay it flat, the shape of the graticule on the flat paper is different from its shape on the spherical model of the earth. The map projection process has distorted the shape of the graticule.

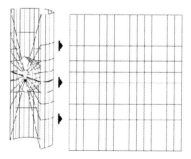

Figure A-3 "Unwrapping" the projected graticules onto a flat surface.

A spheroid can't be flattened to a plane any more easily than a piece of orange peel can be flattened —it will rip. Therefore, representing the earth's surface in two dimensions causes distortion in the shape, area, distance, or direction of the data. Different projections cause different types of distortions (figure A-4). Some projections are designed to minimize the distortion of one or two of the data's characteristics. A projection could maintain the area of a feature, for example, but alter its shape.

Given that all flat maps are distorted to some degree, you can choose from many different map projections (ArcGIS supports more than 300 of them). Each is distinguished by its suitability for representing a particular portion and amount of the earth's surface and by its ability to preserve distance, area, shape, or direction. Some map projections minimize distortion in one property at the expense of another, while others strive to balance the overall distortion. As a mapmaker, you can decide which properties are most important and choose a projection that suits your needs.

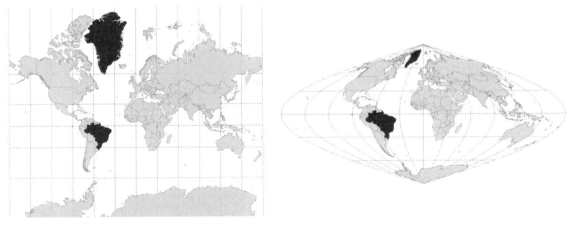

Mercator projection   Sinusoidal projection

Figure A-4   In the Mercator projection, Greenland looks larger than Brazil, although Brazil is four times its size. Because direction is preserved, Brazil correctly appears due south of Greenland. In the sinusoidal projection, the proportional sizes of Greenland and Brazil are correct. Their shapes, however, are distorted: Greenland is too narrow, and Brazil is too wide. (Reprinted from Ormsby et al., 2008.)

Your choice of map projection allows you to control the type of distortion in a map for your area of interest. If you are working with a fairly small area and using an appropriate projection, the effects of distortion are insignificant. If you are working with the whole world, a significant distortion of some spatial property is likely.

If your spatial data references locations with latitude and longitude (e.g., decimal degrees), you can still display the data on your map. ArcMap draws the data by simply treating the latitude-longitude coordinates as planar x,y coordinates. If your map doesn't require a high level of locational accuracy (e.g., you won't be performing queries based on location and distance, or you just want to make a quick map), you might decide not to transform your data to a projected coordinate system. If, however, you need to make precise measurements on your map, you should choose a projected coordinate system. It allows you to measure length, distance, or area, using units of feet, meters, or other linear units.

ArcCatalog and coordinate systems

When you add a layer to a map, both its appearance and the results of measurements and calculations you make depend on its coordinate system. To find a dataset's coordinate system, one must look in its spatial metadata (see chapter 3).

When feature classes with the same coordinate system are added to a data frame in ArcMap, the features in each layer are correctly positioned with respect to each other. If you subsequently add a feature class that has a different coordinate system, ArcMap automatically modifies it to match the others in a process called "on-the-fly projection." This new, temporary projection is applied only within a particular data frame; the dataset's native coordinate system (the one shown in its spatial metadata) does not change.

By default, layers are projected on the fly to the coordinate system of the first layer added to a data frame (even if the layer is later removed). The coordinate system is stored as a property of the data frame and can be changed. You can project all layers in a data frame to any coordinate system ArcMap supports.

To project a layer on the fly, ArcMap uses the information stored in its geographic coordinate system. On-the-fly projection works best when all layers in the map have the same geographic coordinate system (in other words, when they all use the same model of the earth). On-the-fly projections are less mathematically rigorous than permanent projections (which change the native coordinate system of the dataset). If you plan to use datasets in an exacting analysis, you should project them permanently to the same coordinate system with the Projections and Transformations toolset, which are in the Data Management Tools toolbox in ArcToolbox.

Reference

Ormsby, T., E. J. Napoleon, and R. Burke. 2008. *Getting to Know ArcGIS Desktop.* 2nd ed. Redlands, Calif.: ESRI Press.

Appendix B

Data source credits

Chapter 1 data sources include

\ESRIPress\GISHUM\Chapter1\Admin_Boundaries.shp, from ESRI Data & Maps, 2009, courtesy of ArcWorld Supplement

\ESRIPress\GISHUM\Chapter1\Earthsat_150m_Pakistan.jp2, from ESRI Data & Maps, 2009, courtesy of Earth Satellite Corporation

\ESRIPress\GISHUM\Chapter1\Major_Cities.shp, from ESRI Data & Maps, 2009, courtesy of ArcWorld

\ESRIPress\GISHUM\Chapter1\Populated_Places.shp, from ESRI Data & Maps, 2009, courtesy of Digital Chart of the World

Chapter 2 data sources include

\ESRIPress\GISHUM\Chapter2\Districts.shp, created by the author

\ESRIPress\GISHUM\Chapter2\Roads.shp, created by the author

\ESRIPress\GISHUM\Chapter2\SubDistricts.shp, created by the author

\ESRIPress\GISHUM\Chapter2\Sucos.shp, created by the author

\ESRIPress\GISHUM\Chapter2\Towns.shp, created by the author

\ESRIPress\GISHUM\Chapter2\CropForecast2007.dbf, created by the author

\ESRIPress\GISHUM\Chapter2\FoodSupplyData.dbf, created by the author

\ESRIPress\GISHUM\Chapter2\Indonesia.shp, from ESRI Data & Maps, 2009, courtesy of ArcWorld Supplement

\ESRIPress\GISHUM\Chapter2\MaizeSymbol.emf, created by the author

Chapter 3 data sources include

\ESRIPress\GISHUM\Chapter3\Basedata\Airports.shp, ©MapAction 2007

\ESRIPress\GISHUM\Chapter3\Basedata\Cities.shp, from ESRI Data & Maps, 2009, courtesy of ArcWorld

\ESRIPress\GISHUM\Chapter3\Basedata\CommunitySurvey.shp, ©MapAction 2007

\ESRIPress\GISHUM\Chapter3\Basedata\DamagedBridges.shp, ©MapAction 2007

\ESRIPress\GISHUM\Chapter3\EarthSat\Ghana_150m_EarthSat.img, from ESRI Data & Maps, 2009, courtesy of Earth Satellite Corporation

\ESRIPress\GISHUM\Chapter3\Basedata\Ghana.shp, from ESRI Data & Maps, 2009, courtesy of ArcWorld

\ESRIPress\GISHUM\Chapter3\Basedata\HumanImpact.shp, ©MapAction 2007

\ESRIPress\GISHUM\Chapter3\Basedata\InfrastructureDamage.shp, ©MapAction 2007

\ESRIPress\GISHUM\Chapter3\Basedata\Lakes.shp, ©MapAction 2007

\ESRIPress\GISHUM\Chapter3\Basedata\LivestockLosses.shp, ©MapAction 2007

\ESRIPress\GISHUM\Chapter3\Basedata\MODIS_20070106_03.250m.jpg, ©MapAction 2007

\ESRIPress\GISHUM\Chapter3\Basedata\MODIS_20070106_04.250m.jpg, ©MapAction 2007

\ESRIPress\GISHUM\Chapter3\Basedata\MODIS_20070915_03.250m.jpg, ©MapAction 2007

\ESRIPress\GISHUM\Chapter3\Basedata\MODIS_20070915_04.250m.jpg, ©MapAction 2007

\ESRIPress\GISHUM\Chapter3\Basedata\PopulatedPlaces.shp, from ESRI Data & Maps, 2009, courtesy of Digital Chart of the World

\ESRIPress\GISHUM\Chapter3\Basedata\Railways.shp, ©MapAction 2007

\ESRIPress\GISHUM\Chapter3\Basedata\Rivers.shp, from ESRI Data & Maps, 2009, courtesy of ArcWorld

\ESRIPress\GISHUM\Chapter3\Basedata\Regions.shp, from ESRI Data & Maps, 2009, courtesy of ArcWorld Supplement

\ESRIPress\GISHUM\Chapter3\Basedata\Roads.shp, ©MapAction 2007

\ESRIPress\GISHUM\Chapter3\Basedata\UTM Grid.shp, from ESRI Data & Maps, 2009, courtesy of ArcWorld Supplement, NGA

\ESRIPress\GISHUM\Chapter3\Basedata\WestAfricaCountryOutlines.shp, from ESRI Data & Maps, 2009, courtesy of ArcWorld

Chapter 4 data sources include

\ESRIPress\GISHUM\Chapter4\FEWSNET, courtesy of USGS, USAID, FEWS NET

\ESRIPress\GISHUM\Chapter4\Basedata\AffectedCommunities.dbf, ©MapAction 2007

\ESRIPress\GISHUM\Chapter4\Basedata\DamagedSchools.txt, ©MapAction 2007

Chapter 5 data sources include

\ESRIPress\GISHUM\Chapter5\Roads_Export.shp, data simulated by iMMAP. Not for redistribution or use in mapping, navigation, or logistical planning

\ESRIPress\GISHUM\Chapter5\Slope.shp, from ESRI Data & Maps, 2009, courtesy of NASA, NGA, NGA, USGS EROS, ESRI

Chapter 6 data sources include

\ESRIPress\GISHUM\Chapter6\KosovoERW.gdb\VegetatedAreas\Farmland, courtesy of Information Management and Mine Action Programs, Inc. (iMMAP)

\ESRIPress\GISHUM\Chapter6\KosovoERW.gdb\ERWContaminatedAreas\ClusterBombSites, courtesy of Information Management and Mine Action Programs, Inc. (iMMAP)

\ESRIPress\GISHUM\Chapter6\KosovoERW.gdb\Administrative\Districts, courtesy of Information Management and Mine Action Programs, Inc. (iMMAP)

\ESRIPress\GISHUM\Chapter6\KosovoERW.gdb\Administrative\Kosovo, courtesy of Information Management and Mine Action Programs, Inc. (iMMAP)

\ESRIPress\GISHUM\Chapter6\KosovoERW.gdb\Administrative\Municipalities, courtesy of Information Management and Mine Action Programs, Inc. (iMMAP)

\ESRIPress\GISHUM\Chapter6\KosovoERW.gdb\Roads, courtesy of Information Management and Mine Action Programs, Inc. (iMMAP)

\ESRIPress\GISHUM\Chapter6\KosovoERW.gdb\Administrative\Towns, courtesy of Information Management and Mine Action Programs, Inc. (iMMAP)

\ESRIPress\GISHUM\Chapter6\KosovoERW.gdb\ERWContaminatedAreas\UXOLandmineSites, courtesy of Information Management and Mine Action Programs, Inc. (iMMAP)

\ESRIPress\GISHUM\Chapter6\KosovoERW.gdb\VegetatedAreas\Woodland, courtesy of Information Management and Mine Action Programs, Inc. (iMMAP)

Chapter 7 data sources include

\ESRIPress\GISHUM\Chapter7\FDPs.shp, data simulated by iMMAP. Not for redistribution or use in mapping, navigation, or logistical planning

\ESRIPress\GISHUM\Chapter7\Regions.shp, from ESRI Data & Maps, 2009, courtesy of ArcWorld Supplement

\ESRIPress\GISHUM\Chapter7\Warehouses.shp, data simulated by iMMAP. Not for redistribution or use in mapping, navigation, or logistical planning

\ESRIPress\GISHUM\Chapter7\UNSDI-ETH-GDB.mdb, data simulated by iMMAP. Not for redistribution or use in mapping, navigation, or logistical planning

Chapter 8 data sources include

\ESRIPress\GISHUM\Chapter8\Districts.shp, courtesy of ILRI and its GIS Unit

\ESRIPress\GISHUM\Chapter8\Populated_Places.shp, from ESRI Data & Maps, 2009, courtesy of Digital Chart of the World

\ESRIPress\GISHUM\Chapter8\Roads.shp, courtesy of CartONG

\ESRIPress\GISHUM\Chapter8\UXOs.shp, courtesy of CartONG

\ESRIPress\GISHUM\Chapter8\SRTM_Uganda, from ESRI Data & Maps, 2009, courtesy of NASA, NGA, USGS, EROS, ESRI

Chapter 9 data sources include

\ESRIPress\GISHUM\Chapter9\Padibe_Camp.gdb\Administrative\All_Huts, courtesy of CartONG

\ESRIPress\GISHUM\Chapter9\Padibe_Camp.gdb\Services\All_Boreholes, courtesy of CartONG

\ESRIPress\GISHUM\Chapter9\Padibe_Camp.gdb\Services\Latrines, courtesy of CartONG

\ESRIPress\GISHUM\Chapter9\Water_Point.dbf, courtesy of CartONG

\ESRIPress\GISHUM\Chapter9\Padibe_Camp.gdb\Padibe_Image, satellite imagery courtesy of GeoEye

Appendix C

Data license agreement

Important:
Read carefully before opening the sealed media package

Environmental Systems Research Institute Inc. (ESRI) is willing to license the enclosed data and related materials to you only upon the condition that you accept all of the terms and conditions contained in this license agreement. Please read the terms and conditions carefully before opening the sealed media package. By opening the sealed media package, you are indicating your acceptance of the ESRI License Agreement. If you do not agree to the terms and conditions as stated, then ESRI is unwilling to license the data and related materials to you. In such event, you should return the media package with the seal unbroken and all other components to ESRI.

ESRI license agreement

This is a license agreement, and not an agreement for sale, between you (Licensee) and Environmental Systems Research Institute Inc. (ESRI). This ESRI License Agreement (Agreement) gives Licensee certain limited rights to use the data and related materials (Data and Related Materials). All rights not specifically granted in this Agreement are reserved to ESRI and its Licensors.

Reservation of Ownership and Grant of License: ESRI and its Licensors retain exclusive rights, title, and ownership to the copy of the Data and Related Materials licensed under this Agreement and, hereby, grant to Licensee a personal, nonexclusive, nontransferable, royalty-free, worldwide license to use the Data and Related Materials based on the terms and conditions of this Agreement. Licensee agrees to use reasonable effort to protect the Data and Related Materials from unauthorized use, reproduction, distribution, or publication.

Proprietary Rights and Copyright: Licensee acknowledges that the Data and Related Materials are proprietary and confidential property of ESRI and its Licensors and are protected by United States copyright laws and applicable international copyright treaties and/or conventions.

Permitted Uses: Licensee may install the Data and Related Materials onto permanent storage device(s) for Licensee's own internal use.

Licensee may make only one (1) copy of the original Data and Related Materials for archival purposes during the term of this Agreement unless the right to make additional copies is granted to Licensee in writing by ESRI.

Licensee may internally use the Data and Related Materials provided by ESRI for the stated purpose of GIS training and education.

Uses Not Permitted: Licensee shall not sell, rent, lease, sublicense, lend, assign, time-share, or transfer, in whole or in part, or provide unlicensed Third Parties access to the Data and Related Materials or portions of the Data and Related Materials, any updates, or Licensee's rights under this Agreement.

Licensee shall not remove or obscure any copyright or trademark notices of ESRI or its Licensors.

Term and Termination: The license granted to Licensee by this Agreement shall commence upon the acceptance of this Agreement and shall continue until such time that Licensee elects in writing to discontinue use of the Data or Related Materials and terminates this Agreement. The Agreement shall automatically terminate without notice if Licensee fails to comply with any provision of this Agreement. Licensee shall then return to ESRI the Data and Related Materials. The parties hereby agree that all provisions that operate to protect the rights of ESRI and its Licensors shall remain in force should breach occur.

Disclaimer of Warranty: The Data and Related Materials contained herein are provided "as-is," without warranty of any kind, either express or implied, including, but not limited to, the implied warranties of merchantability, fitness for a particular purpose, or noninfringement. ESRI does not warrant that the Data and Related Materials will meet Licensee's needs or expectations, that the use of the Data and Related Materials will be uninterrupted, or that all nonconformities, defects, or errors can or will be corrected. ESRI is not inviting reliance on the Data or Related Materials for commercial planning or analysis purposes, and Licensee should always check actual data.

Data Disclaimer: The Data used herein has been derived from actual spatial or tabular information. In some cases, ESRI has manipulated and applied certain assumptions, analyses, and opinions to the Data solely for educational training purposes. Assumptions, analyses, opinions applied, and actual outcomes may vary. Again, ESRI is not inviting reliance on this Data, and the Licensee should always verify actual Data and exercise their own professional judgment when interpreting any outcomes.

LIMITATION OF LIABILITY: ESRI SHALL NOT BE LIABLE FOR DIRECT, INDIRECT, SPECIAL, INCIDENTAL, OR CONSEQUENTIAL DAMAGES RELATED TO LICENSEE'S USE OF THE DATA AND RELATED MATERIALS, EVEN IF ESRI IS ADVISED OF THE POSSIBILITY OF SUCH DAMAGE.

No Implied Waivers: No failure or delay by ESRI or its Licensors in enforcing any right or remedy under this Agreement shall be construed as a waiver of any future or other exercise of such right or remedy by ESRI or its Licensors.

Order for Precedence: Any conflict between the terms of this Agreement and any FAR, DFAR, purchase order, or other terms shall be resolved in favor of the terms expressed in this Agreement, subject to the government's minimum rights unless agreed otherwise.

Export Regulation: Licensee acknowledges that this Agreement and the performance thereof are subject to compliance with any and all applicable United States laws, regulations, or orders relating to the export of data thereto. Licensee agrees to comply with all laws, regulations, and orders of the United States in regard to any export of such technical data.

Severability: If any provision(s) of this Agreement shall be held to be invalid, illegal, or unenforceable by a court or other tribunal of competent jurisdiction, the validity, legality, and enforceability of the remaining provisions shall not in any way be affected or impaired thereby.

Governing Law: This Agreement, entered into in the County of San Bernardino, shall be construed and enforced in accordance with and be governed by the laws of the United States of America and the State of California without reference to conflict of laws principles. The parties hereby consent to the personal jurisdiction of the courts of this county and waive their rights to change venue.

Entire Agreement: The parties agree that this Agreement constitutes the sole and entire agreement of the parties as to the matter set forth herein and supersedes any previous agreements, understandings, and arrangements between the parties relating hereto.

Appendix D

Installing the data and software

GIS Tutorial for Humanitarian Assistance includes one DVD with exercise data and one DVD with ArcGIS Desktop 9.3.1 (ArcEditor license, single-use, 180-day trial) software. You will find both in the back of this book. Installation of the exercise data DVD requires about 973 MB of hard-disk space. Installation of the ArcGIS Desktop software DVD with extensions requires at least 3.2 GB of hard-disk space. Installation times will vary with your computer's speed and available memory.

If you previously installed data for another tutorial, you cannot simply copy the current data over it. You must uninstall the previous data before you install the exercise data that comes with this book.

If you already have a licensed copy of ArcGIS Desktop 9.3.1 or higher installed on your computer (or accessible through a network), do not install the software DVD. Use your licensed software to do the exercises in this book. If you have an older version of ArcGIS installed on your computer, you must uninstall it before you can install the software DVD that comes with this book.

The exercises in this book work only with ArcGIS Desktop 9.3.1.

Installing the data

Follow these steps to install the exercise data.

1 **Put the data DVD in your computer's DVD drive. A splash screen will appear.**

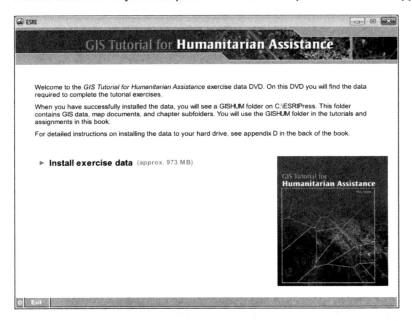

2 **Read the welcome, then click the Install exercise data link. This launches the InstallShield Wizard.**

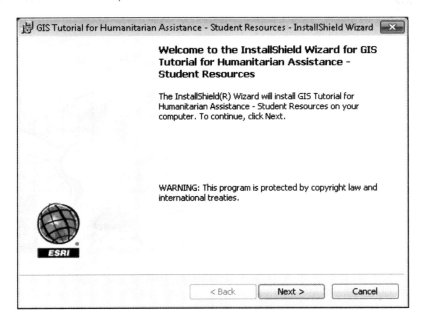

3 **Click Next. Read and accept the license agreement terms, then click Next.**

4 Accept the default installation folder, or click Browse and navigate to the drive or folder location where you want to install the maps and data.

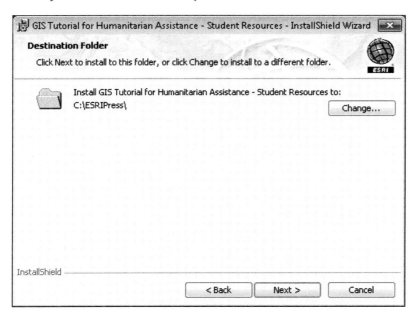

5 Click Next. The installation will take a few moments. When the installation is complete, you will see the following message.

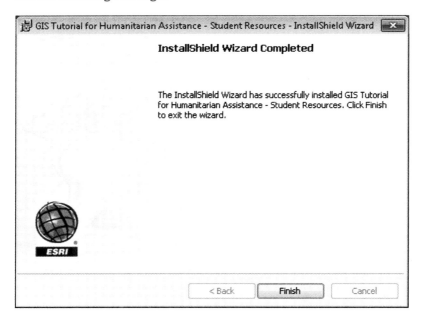

6 Click Finish. The maps and data are installed on your computer in a folder called GISHUM.

Uninstalling the maps and data

To uninstall the maps and data from your computer, open your operating system's control panel, and then double-click the Add/Remove Programs icon. In the Add/Remove Programs dialog box, select the following entry, and then follow the prompts to remove it:

GIS Tutorial For Humanitarian Assistance - Student Resources

Installing the software

The ArcGIS software included on this DVD is intended for educational purposes only. Once installed and registered, the software will run for 180 days. The software cannot be reinstalled nor can the time limit be extended. It is recommended that you uninstall this software when it expires.

Follow these steps to install the software.

1   Put the software DVD in your computer's DVD drive. A splash screen will appear. If your auto-run is disabled, navigate to the contents of the DVD, and then double-click the ESRI.exe file to begin.

2   Click the ArcView installation option. On the Startup window, click Install ArcGIS Desktop. This will launch the Setup wizard.

3   Read the Welcome, then click Next.

4   Read the license agreement. Click "I accept the license agreement," and then click Next.

5   The default installation type is Typical. You must choose the Complete install, which will add extension products that are used in the book. Click the button next to Complete install.

6   Click Next. Accept the default installation folder, or click Browse and navigate to the drive or folder location where you want to install the software. Click next.

7   Accept the default installation folder, or navigate to the drive or folder where you want to install Python, a scripting language used by some ArcGIS geoprocessing functions. (You won't see this panel if you already have Python installed.) Click Next.

8   The installation paths for ArcGIS and Python are confirmed. Click Next. The software will take some time to install on your computer. When the installation is finished, you will see the following message: You must register with ESRI and obtain an authorization for ArcView. The Registration Wizard is also available from Start > Programs > ArcGIS > Desktop Administrator.

9   Click Register Now and follow the registration process. The registration code is located at the bottom of the software DVD jacket in the back of the book.

If you have questions or encounter problems during the installation process, or while using this book, please use the resources listed below. (The ESRI Technical Support Department does not answer questions regarding the ArcGIS software DVD, the GIS *Tutorial for Humanitarian Assistance* data DVD, or the contents of the book itself.)

• To resolve problems with the trial software or with the maps and data, or to report mistakes in the book, send an e-mail to ESRI workbook support at workbook-support@esri.com.

• To stay informed about exercise updates, FAQs, and errata, visit the book's Web page at www.esri.com/esripress.

Uninstalling the software

To uninstall the software from your computer, open your operating system's control panel, and then double-click the Add/Remove Programs icon. In the Add/Remove Programs dialog box, select the following entry, and then follow the prompts to remove it:

ArcGIS Desktop

Related titles from ESRI Press

Analyzing Urban Poverty: GIS for the Developing World

ISBN: 978-1-58948-151-0

Urban Poverty: GIS in the Developing World demonstrates how GIS can be used to improve quality of life in poor urban areas. Chapters cover site analyses of the natural and built environments, visualization of poverty maps, development of appropriate improvement proposals, management of projects, organization of communities, and encouragement of their participation.

Getting to Know ArcGIS Desktop, Second Edition, Updated for ArcGIS 10

ISBN: 978-1-58948-260-9

This workbook introduces principles of GIS as it teaches the mechanics of using ESRI's leading technology. Key concepts are combined with detailed illustrations and step-by-step exercises to acquaint readers with the building blocks of ArcGIS Desktop, including ArcMap, for displaying and querying maps; ArcCatalog, for organizing geographic data; and ModelBuilder, for diagramming and processing solutions to complex spatial analysis problems.

Lining Up Data in ArcGIS: A Guide to Map Projections

ISBN: 978-1-58948-249-4

Lining Up Data in ArcGIS: A Guide to Map Projections is an easy-to-navigate troubleshooting reference for any GIS user with the common problem of data misalignment. Complete with full-color maps and diagrams, this book presents techniques to identify data projections and create custom projections to align data. Formatted for practical use, each chapter can stand alone to address specific issues related to working with coordinate systems.

Designing Geodatabases: Case Studies in GIS Data Modeling

ISBN: 978-1-58948-021-6

Designing Geodatabases outlines five steps for taking a data model through its conceptual, logical, and physical phases—modeling the user's view, defining objects and relationships, selecting geographic representations, matching geodatabase elements, and organizing the geodatabase structure. Several design models for a variety of applications are considered, including addresses and locations, census units and boundaries, stream and river networks, and topography and the basemap.

ESRI Press publishes books about the science, application, and technology of GIS. Ask for these titles at your local bookstore or order by calling 1-800-447-9778. You can also read book descriptions, read reviews, and shop online at www.esri.com/esripress. Outside the United States, visit our Web site at www.esri.com/esripressorders for a full list of book distributors and their territories.